WORKING
ON THE DARK SIDE
OF THE MOON

A fascinating and delightfully personal look at the National Security Agency, the US surveillance giant so secretive it has long been nicknamed "No Such Agency".

> Professor Christopher Moran, Warwick University; author of *Company Confessions: Secrets, Memoirs, and the CIA* and *Classified: Secrecy and the State in Modern Britain.*

A remarkably honest and personal narrative of the otherwise (properly) impenetrable activities at an underappreciated agency, whose talented and hardworking people do a superb job every day to help keep America safe. I have not seen any other unclassified source which puts such a human face on this important work.

> Dr. Lee Neuwirth, former Deputy Director, Institute for Defense Analyses Communications Division; author of *Nothing Personal: The Vietnam War in Princeton, 1965-1975*

Tom Willemain is, like me, an outsider who had the opportunity to spend time inside the world of American intelligence. His reminiscences are not only charming, they are insightful. ... Coming from the outside, Tom portrays the culture in living color. ... Tom is gracious with those who redacted his manuscript, ostensibly for security reasons. Sadly, those redactions are a reminder that the intelligence agencies still don't get it: if they are to be supported by Americans, they need to be much more open in talking about what they do and, especially, about the wonderful people who do it.

> Dr. Gregory Treverton, former Chair, National Intelligence Council and advisor to the Director of National Intelligence; author of *Intelligence for an Age of Terror* and *Dividing Divided States*

WORKING ON THE DARK SIDE OF THE MOON

LIFE INSIDE THE
NATIONAL SECURITY AGENCY

THOMAS REED WILLEMAIN, PH.D.

MCP - MAITLAND

Copyright © 2017 by Thomas Reed Willemain

Mill City Press, Inc.
2301 Lucien Way #415
Maitland, FL 32751
407·339·4217
www.millcitypublishing.com

All rights reserved. No part of this publication may be reproduced, stored in a retrieval system, or transmitted, in any form or by any means, electronic, mechanical, photocopying, recording, or otherwise, without the prior written permission of the author.

ISBN-13: 9781629528724

Printed in the United States of America

"My main message...is the simple words: 'thank you.' I appreciate what you do and the country should appreciate what you do. You are vital to keeping this country safe... Because of your efforts, America's safer today than it was in 2001."

— President Obama to the NSA

"We literally do save lives. I work with people who are the smartest, most ethical in the country, and they care deeply about the Constitution."

— Dr. Deborah Frincke,
NSA Research Director

"I can't think of anything that excites a greater sense of childlike wonder than to be in a country where you are ignorant of almost everything."

— Bill Bryson

Foreword

Sabbaticals Program: The MSP [Mathematical Sciences Program] Sabbatical Program offers mathematicians, statisticians, and computer scientists the unique opportunity to work side by side with other NSA scientists on projects that involve cryptanalysis, coding theory, number theory, discrete mathematics, statistics, probability, and many other subjects. Sabbatical visitors must be U.S. citizens.

During academic year 2007-2008, I was one of two professors selected to be sabbatical visitors at the National Security Agency (NSA) headquarters facility inside Fort Meade, MD. When the sabbatical year was over, I returned during the subsequent summers to continue working, first at NSA and later at an affiliated think tank called the Institute for Defense Analyses Center for Computing Sciences (IDA/CCS).

I did then, and still do now, regard the secret world with a certain amount of awe. Whenever I pass through the fence at NSA without getting shot, I get the same feeling of "Wow, what *I* am doing *here*?" that I got at other times in my life: when I first set foot inside Princeton, then MIT, then Harvard, then Rensselaer Polytechnic Institute, then the Federal Aviation Administration.

The secret world is largely unknown to most Americans. It is, in some ways, like the dark side of the moon. Before we got a good look, the dark side was a mystery and a breeding ground for legends: Aliens surely had bases there, or Russians surely had bases there, or both, or worse. When NSA does appear in the news, it is usually described in a negative way: Here be monsters. There are some prominent books describing NSA in a big picture way (see books by James Bamford and Matthew Aid and books partially focused on NSA by Michael Hayden and Michael Morell). Those books are like lunar orbiters, showing a high-level view of the moonscape. However, there is little information available to the public about the day-to-day life within NSA or supporting parts of the Intelligence Community (IC) like IDA/CCS. This book has moon dust on its boots. It's a view from down in the craters.

This book describes the people who work "inside" and what life felt like to me as a person constantly jumping between my life "outside" as a professor and software entrepreneur and my life "inside" trying to use statistical analysis to keep our and our allies' countries safe. The distinction between inside and outside sits deep in the consciousness of those who work inside. There are relatively few of us who go back and forth between inside and outside, so my perspective on NSA may be usefully unique. Since I have had these parallel lives (academic, entrepreneurial) while "summering" in the IC, readers may also find it interesting to get a peek inside those other worlds too. Accordingly, I often draw contrasts between life in a university or in a software firm and life in the IC.

This account is very personal. I describe my impressions and memories, as best I can recall them, of actual events, places, and people. I have obfuscated places and names by giving them "cover terms" (i.e., aliases), which appear in ALL CAPITALS. ("Obfuscation" is a minor preoccupation in the secret world and anathema in the academic world, illustrating the tensions of a life both inside and outside.) Much of this book is about life "outside the wire"; while these parts do not bear directly on life inside the IC, they show something of what is involved in taking a sabbatical leave at the NSA and may be useful to anybody who would like to follow me inside.

Despite the use of cover terms, the people described here will no doubt recognize themselves and their colleagues. If any were troubled by their depictions, NSA insisted that they had the right to remove themselves from this book. With one exception, they all apparently agreed to let me use them as examples of the talented and dedicated people who make up the Intelligence Community. It has been my privilege to serve with them. They have all been told that the Agency now considers me a "member of the press" (there's a surprise), and they must self-report contact with me if we discuss the Agency. This has probably burned some friendships, and is the only part of writing this memoir that sorrows me.

I have worked hard to give away no secrets. I incurred and acknowledge a lifelong obligation to obtain NSA clearance for any writings concerning my work with the Agency. NSA cleared this book through its prolonged, rigorous and sometimes hostile pre-publication review process. I have left in the parts redacted by the Agency, which appear as black bars like ███████. Most redactions were well-founded, protecting the people who need to work in secret and the topics, sources and methods that must be protected. Some redactions are silly. All would have been interesting to read. I left in the redaction bars to give you some idea of which topics might be "juicy" and how intensely NSA combed through the manuscript. I did not count very carefully, but I estimate 10% to 15% of the words have been blacked out. (For an account of my experience with the prepublication review of this book and the lessons I've drawn from it, see the blog entry at *www.lawfareblog.com/personal-tale-pre-publication-review* .)

Despite this manuscript having survived pre-publication review, the views and opinions expressed in the material are my own and do not reflect those of NSA or IDA/CCS. Nor does NSA's clearance imply that the Agency certifies the factual accuracy of any part of the book. But I do.

Acknowledgements

I have several people to thank:

- My wife, Lucinda, who carried on alone while I disappeared into the secret world, comforted me when what I saw there distressed me, and accepted the loss of income when I was away and when I took leave from my company to write this book. (If life had been different, it is Lucinda who would have been in the secret world, for she would have been a perfect fit.)

- The relative who had a change of heart after threatening to never speak to me again upon learning that I would be taking a sabbatical at NSA, and all those in the family who cheered me on and were even willing to wear over-priced stuff from the NSA gift shop.

- My business partners, Nelson Hartunian and Charles Smart, for understanding my need for one more period of national service.

- Several friends who offered comments about drafts (mostly accusing me of sounding too much like a professor; they suggest that you read this book slowly.)

- Two senior people in the IC who believed me when I said I was writing this book for the right reasons. Many others that I had counted as friends were suspicious or worse, so it made all the difference to hear that Lee and Joe had confidence in the rightness of my motives.

Sabbatical Seduction

Thus begins a strange tale. Every seven years, professors usually have the opportunity to "go away and learn something new". The goal is to come back smarter, with more to offer in the classroom and more research ideas to pursue. My first sabbatical was actually a half-sabbatical (one academic term) spent in my freezing home basement writing up a final report on my research about the creative mystery of what expert mathematical modelers do during the first hour in the life of a model.

My second sabbatical leave was rather more daring, in a professor-ish sort of way. I wanted to do something important in public service that would teach me a lot but also carry with it a substantial risk of failure. This seemed more intriguing than doing something more obviously daring, like taking up skydiving. My answer was the Federal Aviation Administration (FAA).

I had no professional background in aviation. But I had an endorsement from a major figure in air traffic control, Professor Amedeo Odoni, who had served on my PhD dissertation committee at

MIT. With his help, I convinced the FAA to take me on board for the spring 2001 semester (after spending the fall 2000 semester writing research papers and giving seminars at Lancaster University in England). The FAA had no formal sabbatical program and therefore no clue what to do with me, which had its advantages: I could do what I wanted, which was research on topics in aviation that attracted me.

Eventually, I found a productive routine. Tuesdays I would take the Metro to an FAA outpost in McPherson Square, near the White House, to work with Dave Knorr's group on the "Free Flight" problem. Free Flight isn't fully implemented even yet, but it promises to improve efficiency by letting planes fly directly from A to B. Currently, planes fly along invisible highways in the sky, many of which go back to the first air routes established by lighting fires in farmers' fields to guide pilots back in the day. These air routes can be quite roundabout, and they can have traffic jams just like terrestrial highways.

Fridays, I'd take the Metro out to the University of Maryland, to work with Professor Mike Ball's world-class research group on more general problems in aviation operations research. There I got to give seminars, attend seminars, and help mentor Mike's PhD students, so many of whom are now prominent in the field. Besides helping Mike's students, I eventually returned to Rensselaer with topics and datasets useful for courses and graduate students of my own.

The rest of the work week I calculated results in my cubicle while waiting for questions from colleagues (which never came). Saturdays were spent continuing the work in my studio apartment near the Iwo Jima memorial. When I did get out, it was to wander through the nearby Arlington National Cemetery or maybe grab fast food and a movie at the closest mall. One lonely Saturday night I accidentally attended my first "black movie". At first, I found it unsettling to hear folks all over the theatre talking back to the screen. After a while, though, I realized that what was happening was a kind of distributed dialog between the men and the women in the audience. Interesting.

Sundays I would walk over the key bridge to Georgetown University to attend mass in the basement of a building, feeling like an early Christian in the catacombs. I can testify that walking the Key Bridge in winter has its thrills: icy winds (bad), low-flying aircraft zigging and zagging along the Potomac (fun), and strange people who corner you and look like they might toss you over the railing (not fun).

My sabbatical experience with the NSA was hugely different. NSA has a very structured sabbatical program, designed, I think, to do two things: primarily, to seed the academic community with sympathetic allies if not ambassadors, and secondarily to get some work done using fresh minds. I cannot recall when or why applying for this program became a goal for me, but I do know that there were several obstacles, which must have satisfied my need for risk. First, there is an intense (and intensely tedious) background security check that takes the better part of a year, so there are logistical hurdles to participation. Second, NSA has high standards, so applicants get to experience again that anxious feeling they felt back in high school when applying to college. Third, the program is oriented toward mathematicians, which created extra risk for me. I'm an electrical engineer turned applied statistician and software entrepreneur who has no clue about the abstract mathematical fields important to many of the problems of interest to NSA. Fourth, I had to wrestle with the possibility of serious moral risk. Without knowing many details, I had a bad feeling that the Agency had been used improperly by the Bush/Cheney administration. I wondered whether I could work at NSA without complicity, even if the work were only dry statistical analysis. (I was not yet clued in to the kill chain, about which more later).

There was a wonderful woman running the sabbatical program at the time, ADELE. [Note: every name that appears in CAPITALS is a cover name, chosen at random to protect the identity of the person described. Foreign intelligence services would love any assistance in cataloging who's who in the Intelligence Community.] ADELE was a PhD mathematician and a serious runner and bicyclist. More importantly, she was an open and honest correspondent. Over a series of emails, I sounded her out about the program and my doubts about the morality of the Agency's mission. A key factor in my deciding to apply was her assurance that my qualms were not disqualifying. Later, when my wife and I drove down from New York to Maryland to find an apartment for me to use for the year, she met us in a parking lot for breakfast. I was surprised to see bumper stickers on her car supporting "Anybody but Bush" and other causes leaning toward the political left. This surprised me, though I came to see that the Agency is staffed with much more of a cross-section of America than I had imagined. The common denominators are brains and patriotism.

One part of the selection process is a personal interview. Mine was conducted by a ▇▇▇▇▇▇▇▇▇▇▇▇▇▇▇▇▇. [Note: black bars indicate words redacted by NSA in what is called "pre-publication review". For an account of my experience with the prepublication review of this book, see my blog entry at www.lawfareblog.com/personal-tale-prepublication-review]. The interview was administered in my university office. The interview is intended to discover ▇▇▇▇▇▇▇▇▇, criminality, or vulnerability to comprise for reasons pharmaceutical, financial or sexual. One line of questions was about ▇▇▇▇▇▇▇▇▇▇. I admitted to ▇▇▇▇▇▇▇▇▇▇▇▇▇▇▇▇▇▇. He asked ▇▇▇▇▇▇▇▇. The real answer was no, but I foolishly thought back to college and courting days and said "sometimes". Then he surprised me by asking how I acted when drunk. Here was an entirely new question! I foolishly thought, Let's have fun with it. So I answered "I get chatty." He reacted with alarm: ▇▇

At this point, for not the first time in my life, I regretted getting cute with my answer. Not a good start.

But apparently not fatal. Maybe I was rescued by another part of the background check, this one involving interviews. They interviewed my two business partners, my university department head, two friends in town, and random neighbors on my street. Checking with neighbors has since become a suspect approach: Even living in the same house for over 30 years, I still have never even met most of my neighbors, don't know the names of all those I have met, and regularly am surprised when houses sell and new people move in. So I don't think of our street a good information source, but apparently it did me no harm. Those interviewed apparently thought I'd be unlikely to be an Agent of a Foreign Power or a treasonous scoundrel.

Another part of the screening process was completion of a massive questionnaire, the SF-86 form.

> What type of information is requested on a security clearance application? The application form, Standard Form 86—SF86 (Questionnaire for National Security Positions), requires personal identifying data, as well as information regarding citizenship, residence, education, and employment history; family and associates; and foreign connections/travel. Additionally, it asks for information about criminal records, illegal drug involvement, financial delinquencies, mental health counseling, alcohol-related incidents and counseling, military service, prior clearances and investigations, civil court actions, misuse of computer systems, and subversive activities. The number of years of information required on the form varies from question to question—many require 7 years, some require 10 years, and others are not limited to any period of time.

For the NSA sabbatical program, the information went all the way back to grade school. I reported my tonsillectomy in (I guessed) second grade, the names of long-dead teachers, past addresses I could barely remember, and many other tedious (and not very useful) items. Most relevant, in an accidental way, may have been my list of foreign connections. As a professor at a technological university, most of the graduate students in my classes were foreign citizens. And as one-time director of doctoral admissions for my department, I was neck-deep in emails from "interesting places" like Iran and China. (I once admitted a brilliant candidate from Iran. She got a sweetheart deal from the University of Florida that we could not match and enrolled there instead. But that has not stopped dozens of Iranians from emailing me in the years since, just in case I was their ticket to the US. I've often wondered who in the government has had to spend time reviewing all of my email contact with Iran.)

A fun bit of the selection process is a polygraph exam. I suppose that NSA is aware of the research documenting the flaws in polygraphs, but they do serve to intimidate a certain number of people. Besides, what bureaucrat would want to be responsible for ending poly's if somebody bad slipped through? To me, this was going to be interesting. I flew to Baltimore, where I was put me up in a chain hotel near the airport. Looking back, I can't believe how wrong I was about how important this was to NSA. I expected that they had people observing me all the time, and I even did an amateur check for cameras and microphones in the hotel room. I see now that I would not have been worth the effort. The next morning I missed the shuttle bus that was to take a number of us for our polygraphs, but eventually blundered my way to a nondescript office building where the exam would be administered.

The exam itself was interesting. I was ███████████████ which of course drives a professor nuts! We can always say something that we think is more enlightening than ███████████████. Apparently, enlightenment was not the goal. The examiner did note that I had a Ph.D. and warned me against doing what all such animals do: overthinking the questions. Hey, we think it's fun to overthink questions. It is hard not to overthink questions. It's our job to overthink questions. Did I consider myself loyal to the United States? Yes, though not so much to the bits about income inequality, corporate malfeasance, racism, reverse racism, super PACs, and the hated New York Yankees. It's complicated. I guess that still nets out to "yes", but my somewhat mixed feelings may have messed up the results.

I passed. Outwardly, it was uneventful. Inwardly, I thought it would be interesting to do it again. Later people on the Inside told me that ███ because by then ███ In fact, when my clearance ███████████████████████████████████████ later, and they got around to ██ (what's a ████████████████████ among friends?), I was disappointed that they couldn't ██ and limited █████████████████████████████████████ in which I confessed to a few minor security slipups.

One lapse that happened more than once was bringing my cell phone inside from the parking lot. This is potentially a serious problem, and maybe I'm being defensive to call it a "minor security slipup". Perhaps most readers can sympathize, as most of us have become one with our phones and feel naked if their weight isn't subconsciously present. But generally speaking, bad people can take control of a phone, so phones should stay outside, even the old fashioned "flip phone" I use. (I may be one of the few remaining Americans who will not use a smart phone. Thanks to work I've done while inside, I've learned to call smart phones "████████████████████████", using that excuse to be contrarian on the subject.)

A definitely less serious lapse involved accidentally bringing a CD inside. The CD in question was inside a thick textbook on operations research that I wanted to use as a reference. The book, like many, had an "Instructor's CD" pasted inside the back cover. I was not even aware of the CD until I opened the book in the workspace that I will call the DARKROOM. I have no doubt that this particular CD was innocent, but CD's are a classic vector for computer viruses, and NSA is rightly paranoid about viruses. (When the Agency reviewed this book to confirm that it contained no security lapses, it insisted on reviewing a paper version, not a version on CD.) When I finally discovered the illicit CD, I dutifully turned myself in. The "sheriff" was ALEXANDER, a former Ivy League professor who had this added security duty on top of his other work, much of which involved mentoring and protecting all the Math Research interns (see below). Since ALEXANDER was such a friendly guy, I got worried when he got worried. He interrogated me about the sorry history of this episode, seized the CD, and actually had it melted down. I felt horrible, of course.

But then ALEXANDER took pity and told me about his own rookie mistake. As a newbie, he had mistakenly ███████████████████████ ████████████████ on some exotic topic in ███████████████████. While ████████ in his ████████ it suddenly hit him that he had committed ███████████████████████████. Then he inadvertently ████████████ In a panic, he not only ████████████████████, he ███████████████████████████ (He stopped short of ███████████████ around his ████████████ Roman style.) Then he called ███████ and ████████████████████. At that point, the ████████████████████████ for ████████████ the ████████ they needed to ████████████ the incident. I appreciated his successful effort to make me feel less bad about my goof.

My main problem that day in the nondescript office building was not the polygraph exam but what came next. They needed to get digital images of all my fingerprints. It turned out that some of my fingers have very indistinct prints. The guy working the machine got more and more frustrated and let me know it was all my fault. Imagine my indignation. I kept my mouth shut and got out with another step in the selection process checked off.

Passing a polygraph test is not nothing (I'm now a Certified Loyal American), but it's also apparently not necessary. The other sabbatical visitor who entered with me, ANDREW, was a math professor and one of the world's straightest shooters. Yet ANDREW actually failed more than one polygraph before they gave up and let him in. In ANDREW's case, taking the poly involved getting up well before dawn, driving hours to a decent sized Midwestern city, flying to

Baltimore, taking the poly, then flying back the same night to be ready to teach the next day. Not convenient, not cheap for the taxpayer, and not useful.

ANDREW is an athlete. He swims, he bikes to work, he has the low metabolic rate of an estivating frog. One time, ANDREW warned the examiner that his ███████████ was ██████████████ and ███████. The examiner ████████████████████ and screamed that ANDREW was ██████████████████████████. (Some examiners are very grouchy people. It might be the job, but it also might sometimes be a tactic. My own examiner was unremarkable – all business, no attitude.)

Another time, ANDREW used his Ph.D. to overthink a question. He was asked if he'd ever been arrested. Well, yes, but by mistake: Is that a yes or a no? It was a birthday; he'd turned the legal drinking age in his state. His buddies took him to a bar. He imbibed probably half of one beer, probably the first and last in his life (Question: Does ANDREW drink? Yes or no?). Then he left early. But on the way out he walked where a waitress had earlier dropped a glass, and his sneakers picked up some shards. Then he drove home the back way, along an isolated rural road. By unlucky coincidence, a house on that road had been broken into that very night. The police saw ANDREW's car, stopped him, somehow detected the glass in the soles of his sneakers, and arrested ANDREW for obviously having glass from the break-in stuck in his shoes. ANDREW was actually in a jail cell for a while. Therefore, ANDREW the Overthinking Ph.D. flunked his polygraph.

It's a good thing the screeners just gave up on ANDREW and let him in even though he tends to flunk polygraph exams. It turned out that ANDREW soon left academia and stayed with the Agency as a permanent employee. ANDREW brings a powerful combination of math and physics knowledge to bear on an important target. As a bonus, his very talented son CALVIN also signed up.

The final stage in the selection process required presentation of a seminar to prove one's intellectual worth. Professors love to give seminars, but an NSA seminar is a little different. First, I had to prepare my PowerPoint slides well ahead of time and send them to Ft. Meade early so they could be checked for malware. When I arrived at NSA, I was carefully observed by stern people with machine guns, entered one of the Visitors Centers, and was escorted into the building where I would speak. At this point, the contrast between "outside" and "inside" became very obvious. Everywhere I went, I was preceded by flashing red ceiling lights warning everyone that an "uncleared" soul was afoot. There is a small distance between "uncleared" and "unclean". It's easy to feel like a leper when, as you walk along, doors close, conversations cease, and curious eye follow you along.

I can't remember much about the seminar itself. I do remember being abandoned to try to work out the projection system, having a large, interested and active audience, discussing topics in time series analysis (my thing), and trying my best to behave well.

Apparently, it was enough. I was in. I was now a member of Mathematics Research, a key part of the Research Directorate. Mutual seduction concluded.

Breaking In:
Fear and Trembling in
the DARKROOM

That first transition from "outside" to "inside" was not easy, and it has continued to be difficult over the years. My wife remained at home during the sabbatical year, and loneliness set in immediately. I flew home for the weekend after just one week, then again after two more weeks, then again after four more weeks. Eventually, I accommodated to living alone with an air bed and furniture made out of cardboard boxes.

My very first day on the job was generally exciting but ended with moments of angst and dread. I was assigned to an overcrowded office in a cold, dingy basement of the ███████████████████ ███████ Building. The sign on the door proclaimed it to be the DARKROOM. The original DARKROOM was a proto-NSA set up in the World War I era.

Inside, I was assigned a temporary seat near the door amidst a few "interns". I soon learned that at NSA's Research Directorate, interns are not callow and clueless novices. No, interns are people who have been plucked from the top tier of their peer group. All those I met that day already had Ph.D.'s.

Interns participate in three year development programs, during which they take classes, whose homework involves real-world problems, and rotate through a few six-to-nine month long rotations in operating offices to help them decide where to land when they are fully fledged. (To my secret delight, much of the course work the interns were made to take was designed to convert theoretical, abstract mathematicians into statisticians.) These development programs are the mathematical equivalent of medical internships and residencies.

One intern in our space, CALEB, was a kind of brighter twin of me, in that he had graduated from Princeton and then done his doctoral work at MIT. One big difference: he is a genius, and I was already on a downward path of cognitive decline.

(I found it interesting that the vetting process had missed that ominous fact, though my investigators had permission to access all my medical records, including the one that recorded my diagnosis of "mild cognitive impairment", which is often a precursor to Alzheimer's. One day a few years earlier, one of my doctoral students had started our weekly meeting very bluntly, telling me that she would review everything we did the previous week since she knew I would not remember any of it. I was stunned. Neuro-psychological testing then revealed that I was scoring in the 16th percentile on some cognitive tests, that I had no ability to remember faces or phrases, could not solve simple puzzles, etc. I was told that there was no way I could continue my career as a professor in an engineering school. Not a good day. Off to a neurologist specializing in Alzheimer's and other dementias, who put me on Aricept and won me at least another decade of useful work.)

A second intern in the DARKROOM was a woman who was planning a wedding that would take place on the same day as my own daughter's wedding: too much synchronicity! How had my life gotten entangled with the lives of these strangers? Was there a quantum computer gone wild someplace down here, where NSA could plausibly also be hiding dead aliens? Her fiancé was also at NSA. I have no data on the subject of the romantic lives of NSA people, but my impression is that inter-Agency dating was not unusual. Certainly, it is simpler. (Remember that problem with getting chatty after two beers?)

The third intern was a wild man who was one of two people I know in the secret world who took to wearing those crazy sneakers with toes on them, which I find repulsive. You know, the ones that look like goofy fabric feet? A few years later, post Snowden, this intern was no longer an intern and was one of the young hotshots paraded out for a story aired on "Sixty Minutes". The whiz kids being interviewed came across as super-smart and super-arrogant. The camera did not show their footwear, so they did not also come across as super-weird. (I found it interesting that all the Old Hands stayed invisible behind the locked door in the background bearing the sign "DARKROOM" –this was not the authentic, dingy DARKROOM but the new sunny suburban edition, about which more later.)

Once installed at a desk, I was introduced around. First stop was the Technical Director ("TD"), COLIN. COLIN was recently diverted from retiring and promoted to a very high position in the Research Directorate. COLIN is a poster boy for technical leadership. If he were in baseball, he would have been a player, a coach, and a general manager. As a player, he was prolific in his own technical output, frequently winning coveted prizes for his work. (The secret world has its own secret prizes and awards, including cash awards, to recognize achievements whose names cannot be spoken. I remember everybody gathering to watch a few stars receive praise and bonus checks for outstanding work. Each individual award speech ended with a statement that it was not permissible to describe whatever brilliant accomplishment was being recognized – even to us. There are secrets within secrets in the secret world.) As a coach, COLIN lavished attention on interns, always sharing co-authorship of the resulting classified reports (would that all professors were so minded). As a manager, COLIN established strong joint efforts with two national labs in the ███████████, which not only tapped into some extra ███████ but helped keep ██████ as some of their other ██████████████████ COLIN paid the price by spending way too much time on airplanes visiting the labs, but he compensated by haunting Yosemite National Park. At this point, every bear and chipmunk in Yosemite must be on a first name basis with COLIN and his vivacious wife.

Like everyone else in the DARKROOM, COLIN had his quirks. He was an inveterate punster, and I often had trouble keeping up, even when the puns were safely inside emails and could be appreciated slowly. One of his colleagues, GEORGE, suffered from the same genetic defect. Putting both of them in the room at the same time was like mixing two explosive chemicals. During technical seminars, it was cognitive overload to absorb the real stuff while keeping tabs on the puns. Imagine a conference room stuff to the gills with big brains, whiteboards covered in juicy equations, some earnest PhD expounding about some exotic topic. Then a crazy call-and-response breaks out across two corners of the room, disrupting the math but funny as hell.

Talking to COLIN was usually awkward for me, partly because of the puns, partly because he generously assumed that I had the math background to understand everything he was saying. I did not. But he was kind and supportive, pointed me to good projects, and gave me a benchmark for judging the quality of technical leadership.

People like COLIN are a national asset and good counter-examples to the common belief that all public sector employees are lazy clock-watchers. The early- and mid-career public servants I taught at Harvard's Kennedy School of Government were bright, hardworking and sincere. Dave Knorr, COLIN's counterpart during my prior sabbatical at the FAA, is a model public employee and a great group leader. My colleagues at the NSA and at IDA/CCS did not come close to fitting the clock-watcher stereotype. The other half of that belief is that people in the private sector are hard-charging dynamos. Since starting a software company in the early 1980's, I have seen enough zombies among my private sector customers to conclude that the "invisible hand" can be pretty numb at times. I believe there is as much drone-osity in the private sector as in the public. Simple generalizations are so easy and so deceptive: private sector good, NSA evil, etc.

COLIN introduced me around the DARKROOM and gave me a too-quick review of my responsibilities regarding secrecy. There was an ancient safe, and I was supposed to be able to open it and store classified documents in it — but I didn't really get to learn how to open it. (Years later, somebody else opened it for me and I retrieved a yellowed, typed document from the 1950's that was useful in one of my research projects. It is quite common for Math Research work to lay dormant for many years before proving useful. That fact must be comforting to abstract mathematicians who can ease any qualms about the utility of their work by telling themselves it will eventually be valuable. Then again, if they had such qualms, they probably wouldn't be abstract mathematicians.)

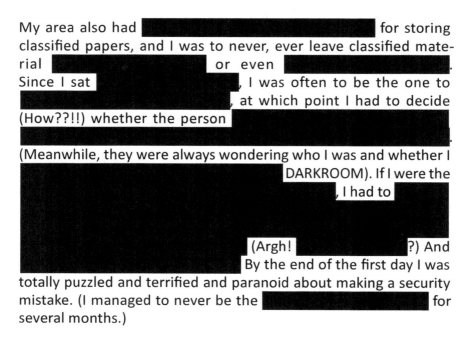

My area also had ███████████████ for storing classified papers, and I was to never, ever leave classified material ███████████ or even ███████████. Since I sat ███████████, I was often to be the one to ███████████████████, at which point I had to decide (How??!!) whether the person ███████████████████████. (Meanwhile, they were always wondering who I was and whether I ████████████████ DARKROOM). If I were the ██████████, I had to ████████████████████████████████ (Argh! ███████?) And ████████. By the end of the first day I was totally puzzled and terrified and paranoid about making a security mistake. (I managed to never be the ███████████████ for several months.)

Then I met GRAHAM. This was consequential, because he soon taught me much about how to live in the secret world, and we went on to write several interesting joint papers and develop a personal friendship on the outside. It was also consequential immediately, because I had my first security insecurity. GRAHAM immediately helped me with job #1, which was to find a project to work on. Since my computer access was not yet up, he gave me a printout of a Top Secret report.

I read it with interest. Somehow, with all the "Top Secret" markings on it, there was a strange electricity in the paper. I'd never thought about the particular topic discussed, but I could skim through and get the drift. On a second reading, I could follow the technical argument. I began to realize that I could do this kind of work. There was a spark of confidence. It was the end of Day 1, and I had hope.

Then I had a kind of fear that was new to me. I looked up and found, to my horror, that it really was the end of Day 1, and GRAHAM had gone home! So, apparently, had everybody else! The next moments are burned into my memory. I thought, "My fingerprints are on this document. I can't leave it out, but I don't know how to lock it away. I'm going to get tossed out for mishandling classified information on my first day at work."

It's understatement to say it was a bad moment. To my great relief, COLIN was actually still around, and he secured the report for me. But the panic and dismay of that moment was great enough to trigger a series of similar bad moments.

In fact, for the next few weeks, I had nightmares about security issues. I remember one in particular. I dreamed I was living in an apartment with paper thin walls. On the other side of the walls prowled a fierce tiger. If I were to make a security error, the tiger would tear through the walls and rip me to shreds. It sounds peculiar at this remove, but the dread behind those nightmares was a burden. Other professors were shaping their sabbaticals into tourist extravaganzas in exotic locales, mixed with well-paid guest lectures and long boozy conversations in which they could say anything they wanted about anything to anybody. There I was living in dread within my sabbatical in sunny Odenton, MD, an Army town with the usual share of bars, tattoo parlors, and liquor stores. The best I could say about my choice compared to my fellow professors' was that I was expanding my horizons in an unconventional way.

The physical work environment in the ▮▮▮▮▮ of the ▮▮▮▮▮▮▮▮ was anything but glamorous. The space was very crowded. Once I was removed from the intern pen, I was squeezed into a desk at the far corner of the far end of the DARKROOM. While we had very little floor space, the ceilings were majestically high, and I sometimes wondered if there was a way to use that space (a hammock?), especially since all the warm air floated up there, leaving the space around my desk rather "invigorating". The desks and filing cabinets were ancient and the chairs creaky, but the computers more than respectable. I was given an ▮▮▮▮▮ and ▮▮▮▮▮▮▮▮▮▮ to keep handy in case of an attack; that was not a good moment for both of the obvious reasons: that we might need it, and that it was not in good repair.

We had one scheduled evacuation drill during my sabbatical, which did not seem to go either well or badly. I would never have found my way out of the labyrinth on my own, but I followed some other geeks to the sunlight. Once outside, we didn't revel in the fresh air and sunlight. Instead, we fretted about the time we were wasting milling around when we could have been computing something.

We also had at least two instances ▮▮▮▮▮▮▮▮▮▮▮▮▮▮▮▮

a ▮▮▮▮▮▮▮▮▮▮▮▮▮▮▮▮▮ The DARKROOM had nobody panicked, but ▮▮▮▮

There were three desks in our room. The first desk sat in the doorway and was visible from the entrance to the DARKROOM suite. This space was called the "Ejection Seat" because the occupant was usually quickly shuttled somewhere else. To me, the Ejection Seat was a bonus resource. Because the most recent occupant had already been ejected, the workstation sat idle. Therefore, I got to run "embarrassingly parallel" computing jobs on the Ejection Seat workstation while simultaneously running other instances on my own workstation. (The technical term means that it is very simple to run more than one computer job at once, using no cleverness whatsoever. Parallel computing can get very complicated, but even I can understand embarrassingly parallel problems. They are the best!)

Each desk had a computer that had been named after someone famous. I embarrassed myself early by asking why everybody else's computer was named after a famous mathematician but mine was named after a notoriously sexy actress. Maybe my basic instinct was to act to type as an engineer plunked down in a nest of mathematicians.

The other great physical resource in our room was a very long and very ancient slate blackboard, perhaps the only one remaining in Math Research (and later lovingly transported to Laurel, MD when all of the Research Directorate was bumped off campus to a suburban office park) and some colored chalk. One of my first acts of secret public service to the United States, on Day 2, was to wash the blackboard, clean the erasers, replenish the chalk supply, and use an antique vacuum cleaner to tidy up afterwards. (For a long time, I had a fear that my janitorial contribution would be my most significant. Looking back, it was probably somewhere in the middle of the top ten.)

The greatest feature of my room was that the middle desk was occupied by a truly admirable NSA researcher, HENRY. Whether he was willing or not, I determined to make him my "rabbi". I grabbed him in a figurative bear hug and tried to squeeze out of him every bit of lore, wisdom and technique. HENRY was unusual in several ways, including being second generation NSA, his father having been part of the Cold War contingent. He was incredibly generous with his time, whether it was filling me in on the invisible mores of the place, explaining technical matters (e.g., ▮▮▮▮▮▮▮▮▮▮) that everybody else knew and assumed that I must know too because, after all, I was sitting in the DARKROOM, and advising me on the technical aspects of my own research projects. He had a bad habit (bad for him, good for me) of declaring "There's no time like the present" and dropping whatever he was doing to answer my questions. His generosity extended well beyond me. During his long career, he had mentored many interns and continued to keep track of them and their careers. HENRY was also helpful in more personal ways: he showed me how to get to the best candy and soda machines, and he and his wife were kind enough to have me to their house for dinner at a time when the loneliness of being a "geographic bachelor" was getting intolerable. Because I had a different arrangement, being a sabbatical visitor, nobody tried to impose work/life balance on me. So I usually worked seven days a week, once I had learned how to open and close the DARKROOM by myself. Invariably, HENRY was there on a Sunday morning, taking care of the non-fun part of his job, i.e., paperwork.

The Walls and the Halls: NSA Outside the DARKROOM

The physical environment outside the DARKROOM was no more luxurious than within it. Located in the basement of a building where computers (which love cold) were more important than people (who usually do not), the corridors were quite chilly (cold air settles downward). The corridors were also quite dingy, painted "government gray" and poorly lit. One happy day I wandered down, anxious as usual about whether I could really crack the problem I was working, when something struck me as odd, even alien. It took a while, but I finally realized that somebody had actually painted the walls with lighter colors. Later, there was another strange day when something again seemed very different. It was: Somebody had installed actual lights in the ceiling, and it was easy to see where I was going. Not quite sunshine, but a definite mood enhancer. Less dungeon, more drab government office building.

To me, the corridors around the DARKROOM were quite frustrating, because along each one were many mysterious and tantalizing doors marked by mysterious and tantalizing signs. As an inveterate academic, I itched to go through each one and ask everybody "What are you working on?" There were doors for ▓▓▓▓▓▓▓▓ linguistics (How cool is that? I sat with those types of people as a grad student at MIT. Let's talk Chomsky!), for ▓▓▓▓▓▓ (mentioned prominently in the ▓▓▓▓▓▓ revelations and one of the golden-boy projects due to its contributions to the wars), and for things I just couldn't suss out in the least. Of course, curiosity is not the same as need-to-know, so the doors never opened for me. It is difficult for a life-long professor to stifle curiosity.

The rest of the NSA campus at Fort Meade (referred to also as NSAW, "NSA Washington", to distinguish it from facilities in Georgia, Texas, Colorado, Hawaii, and Utah) was equally overwhelming and mysterious. I can't prove it, but I'm willing to bet that NSAW has the largest parking lot in the country, especially if we narrow the category to include only monstrously large lots without a single empty space. Once on a hot summer's day I decided to walk to the "Big Four" building (the ones on the front cover of this book) rather than take the shuttle bus; the heat and the distance across the parking lots almost did me in. (One day the thermometer in my old Corolla read 105 degrees.) Big shots get to park near buildings, but the rest get to take their chances in Timbuktu. Coming in to the building known as "Ops 1", one encounters what may still be the longest corridor in the country (dwarfing the famous "Infinite Corridor" at MIT — but at least the Infinite Corridor features an occasional sunset on the west end). I would invariably get lost if I had to go to another building, though I could find the main (Friedman) auditorium and the main cafeteria (run by Sodexo, the same outfit that messed up the food at the Rensselaer Student Union – is there no escape?). I would still get lost now, assuming I could even find a parking space when I drove to the Fort. I pity the poor spy who manages to somehow get into the buildings; pretty soon, he will be begging for rescue, and maybe for better food.

Another place I can find in the Big Four is the NSA gift shop (just follow the country's longest corridor and keep looking for the one door that isn't locked). At first, it seemed strange that NSA would allow employees to buy hats, golf shirts, jackets, shot glasses, mouse pads, laser pointers, and what not, all emblazoned with the NSA logo. I indulged a few times, both for gifts for my relatives but also for myself. Only once has anybody remarked on my NSA windbreaker (he accused me of a security violation during the middle of Catholic mass), and nobody yet has thrown tomatoes at me in public. Not surprisingly, the stuff in the NSA gift shop is overpriced, but at least some of the money goes to the employee welfare fund, so that's ok. What is most strange, to me, is that every single thing in there is made in China. That sounds like a scandal waiting to happen: Can't we find any American vendors? You can be sure that I checked my Chinese-made NSA laser pointer very carefully before using it in secret spaces. If there are any bugs in that device, they eluded me.

But then, I never took the course on bugs. Instead, I took the short version of the very interesting course on "Denial and Deception". That was a wicked pleasure for a professor, who is supposed to be in the Truth Business. I love to brag that I'm a graduate of that course. And I kinda like to say "I can neither confirm nor deny" whenever it's even slightly appropriate. It never hurts to practice. Do I actually practice? Hey, I can neither confirm nor deny.

The last time I visited the NSA gift shop, I noticed that it is keeping up with the times. It is now also the Cyber Command gift shop. The Cyber Command merchandise is pushing out the NSA merchandise. Commerce imitates life, since in the years since Cyber Command was established, it has seized many choice bits of Big Four real estate, displacing such gems as the main NSA library, with its wonderful maps showing all the ▬▬▬▬▬▬▬▬▬▬▬▬ on the planet. Plus which, some of the Cyber Command merchandise is a way cooler than a lot of the NSA merchandise. Since DIRNSA (Director, NSA) is also head of Cyber Command, I'm sure he sees no problem: A man who is "dual hatted" has no standing to object to a gift shop selling two kinds of hats, right? (As I write this, the question of separating Cyber Command from NSA is in the news. I wonder if policy makers are factoring in the economies of scale attendant on having a single gift shop for both.)

One final comment on the walls of NSA. With all that square footage available, they also serve as billboards of a sort. There are some impressively artful patriotic panels, and some cool depictions of the idealized environments of the various NSA locations (Georgia, Texas, Hawaii, Colorado). More seriously, there are what might be called instructional messages. Three especially caught my attention when I arrived. One informed us that "MySpace is their space." By the time I read those posters, MySpace was so yesterday as to make the poster kind of a joke. But the message was still valid. Too many people, in both the military and civilian sides of NSA and in other DoD components, were posting personal and work-related information on social media. It was safe to say that this information was vacuumed up by the other side(s). I took the message to heart and have consistently refused to use Facebook, Twitter, and other social media. My excuse is that "I'm not allowed."

My stronger reason for not using social media is that I'm not interested, and perhaps that I'd like to enjoy the distinction of being the only American who has never used social media. I find it attractive to have a small contrarian streak and to enjoy being "against" things. (Besides social media, for which I have this official excuse, I oppose any coffee and all pets.) Smartphones too: I call them "███████████", since I helped ███████ the ███████████ and those who ███████ are less ███████████ for ███████████████████████, so I can fabricate another "official" excuse to be contrarian. The Grumpy Old Man role is not without its delights.

A second kind of poster on the NSA walls carried the message that "One man's trash is another man's treasure." Which brings us to the subject of burn bags. Security required that any printed matter that is not under approved lock and key must be burned. Most of the career NSA people have long ago gone nearly paperless, but I have not been able to free myself of the need to read and mark up complex technical material in hardcopy form. So in every office in which I have parked myself, I have been the champion generator of classified waste. There are rules for proper disposal of one man's trash:

approved heavy-duty paper bags must be used, must be double-stapled shut, and must be labeled with the room number from which they sprung.

At NSAW, trash day was a big event for me. Since we did not have a secretary in the DARKROOM to handle the trash on a big push cart, I got to go to the loading dock myself and manfully chuck my own bags of classified waste into the back of the big trash truck. The truck looks just like the ones that come through my neighborhood at home, and NSA cleverly turns the burning of trash into the generation of electricity. In that sense, I should have gotten some sort of commendation for being the "greenest" researcher (in both senses of the word). At IDA/CCS (see below), we left the bags ███████████ for ████████████████████████████████ If I let myself, I could imagine that this policy let a team of highly trained people pore through my discarded genius ideas, salvaging those that mortals could understand. (The Walter Mitty syndrome takes many forms.) It is not uncommon for me to generate three, four, or five wrong ideas before finding one that works, but this doesn't bother me anymore. Nor am I concerned that this makes me the biggest trash generator in my office. Everybody has to be best at something.

The third kind of wall poster that caught my eye seemed most ominous. Their common theme was: "If you feel like you are on the verge of a meltdown, call the following number." That was scary. I had never been in an environment where breakdown was so openly acknowledged. I believe that the operational side of NSA was more at risk for breakdown than the Research Directorate, but I can acknowledge that the nature of the work, and the general atmosphere of threat (see "NSA Daily" below) could even impact the quiet world of Math Research. In fact, recently one █████████████████████████ ██████████████████████████████████████. After this event, the ██ ████████████████████████████████████ and offered ███████████████ to all those who ████████████ This incident might have been ███████████████████████, but it is easy to imagine that ██████████████████████ people, often working to support ███████ in ███████, and perhaps having ███████ in a ███████ area on ███████ might ███████ NSA enforces work/life balance by not permitting staff to work more than a normal work week (except during emergencies, like 9/11). I presume this keeps psychological meltdowns to a tolerable few.

As an academic with a side job of owning a software company, the idea that anybody would not work a seven day week seemed unrealistic and rather stupid to me. I recall that the very day I was to sign my retirement papers at Rensselaer, the Provost insisted that we meet anyway to discuss my "conflict of commitment". While some high-minded colleagues refused to fill in the annual forms concerning conflict of interest and conflict of commitment, I had dutifully reported how many hours I spent working on my outside business. It turned out that the university was defining a work week as having five days, while everybody knows it has seven.

During my sabbatical year, I spent at least part of every weekend in the DARKROOM. Missing my family, and unwilling to enjoy myself without including my wife, I felt like I had no choice. But sometimes the trip into the office on a Saturday or Sunday afternoon felt painfully lonely. Add to that my own anxiety about producing any tangible results, and the result was that my sabbatical year was high stress – enough that those posters about breaking down always got my attention. Later, when I worked at IDA/CCS during summers, I was subject to enforced work/life balance. My protests and pleas for an exception were rejected. That was not terrible: there was always more software to make on weekends so I could indulge my work fetish, but it was irritating to sit on an idea until Monday rolled around again.

More Academic than the Academics

As a professor I am steeped in the tradition of frequent seminars and the imperative to create new knowledge. I viewed it as a bargain with society: we would be free to play, and society would ultimately benefit. In principle, this means that university departments must be open to new ideas, must constantly upgrade their knowledge, and must work hard to ensure that the knowledge gets passed on to new minds. In practice I have seen the crush of business in the university impede the development and exchange of new ideas. I have seen departmental seminars devolve to become recruiting and propaganda vehicles. I have seen supposedly eager doctoral students sit mute and never question a guest speaker. I have rarely seen seminars in which unproven, experimental ideas are discussed energetically.

Surprisingly, what I found in my corner of the NSA was a much closer approximation to the academic ideal. I was amazed at the level of intellectual activity in the Math Research organization. Not only the interns but also senior staff would enroll in specialized short courses. There is a standing weekly statistics seminar, featuring both internal and external speakers. There is a weekly "statistical scientists" meeting at which anyone could present a work in progress and get help from the audience. Whenever someone went to an outside conference, they would report back what they learned to their peers. There is an annual conference called "Graph Fest", to which foreign partners are invited, that reports the latest results in applied graph theory. There are annual conferences operated ▓▓▓▓ which ▓▓▓▓ among the ▓▓▓▓ one such is "▓▓▓▓", which reports the latest developments in ▓▓▓▓ It is considered acceptable for a researcher to spend a week or two reading books and journal articles and "just" thinking, especially since he or she will often give a talk about what was learned. In many a modern American university, this behavior would be considered unproductive by trustees, presidents, provosts, associate provosts, assistant provosts, deans, associate deans, assistant deans, department heads, assistant department heads... and any colleagues hoping to become one of the above.

All this stimulating intellectual activity does come with some restrictions generated by the need for secrecy. As an academic, one of my standard modes of greeting to a colleague would be to ask "What are you thinking about?" (A legacy from Professor Aaron Fleischer, one of my senior colleagues when I was on the MIT faculty, who would always greet me with "Tom, tell me everything!") One of my problems with transitioning from outside to inside was that I had a bad habit of wanting to ask that same forbidden question on the inside. Especially at the beginning of the sabbatical year, this habit created a problem. It was especially bad if I really forgot myself and ask the question out loud in a public space like a corridor. Indeed, it is generally not a good idea to ask such a question even behind locked doors. Not everybody has the necessary clearances and, even if they do, they may still not have the need-to-know.

Everyone I worked with had at least a Top Secret security clearance. But since the Snowden affair, newspapers have reported that over one million Americans have Top Secret clearances. There exist "higher" clearances, and we all had those too. Nevertheless, it is still not OK to talk to just anyone about just anything, because there are further restrictions called ████████████████. There is sensitive compartmented information (SCI) and beyond that there is ████████████████████████████████████. I hold three ████-level clearances (a low number compared to my colleagues), yet I have attended seminars at NSA in which the first half involved knowledge of one ████████████, which we all possessed, but then I and others had to leave the room when the discussion touched on another ████████████, which we did not possess. (I thought it slightly slipshod that I was left to self-designate as ineligible to listen; could I have gotten away with staying in place if I had just mustered a confident smile?) The key concept is "need-to-know". Even with high security clearances, "need-to-know means no". (After 9-11, there was an effort made to replace need-to-know with need-to-share, but I didn't see evidence that the switch ever caught on.)

Despite the secrecy restrictions, NSA's Math Research group was in many ways a dream environment for a professor on sabbatical. Apparently, it was also a hospitable environment for ex-professors. I was struck by how many of my colleagues had been academics in an earlier life. I was also struck by how many of them were bitter about their years in academia.

My own academic career has not been without difficulties. I (foolishly?) left my first position on the MIT faculty in disgust over the way a contentious incident was handled by my department. I left my second position on the Harvard faculty partly due to a disagreement with the dean and partly because I was in an intellectual rut and wanted to try something different, more instinctive and less intellectual, which led to my co-founding Smart Software with two friends, Nelson Hartunian and Charlie Smart. I retired from my third academic position on the faculty of Rensselaer Polytechnic Institute with some sadness about the degradation of the academic environment after a new and famously despotic president adopted a corporate model of the university. Along the way, I have had great deans, forgettable deans, and despicable deans; inspiring students and disappointing students; admirable colleagues and forgettable colleagues. But overall, the 40 year span of my academic career felt productive, stimulating, useful, and satisfying.

Not so for all of my colleagues at NSA. More than one told me stories of an academic experience unlike any I had seen myself. For instance, ANDREW used his sabbatical as a means of escaping a Midwestern state university where his department was focused more on graduating primary and secondary school math teachers then on advancing mathematics itself, where the university administration was extremely political, where only a small percentage of the students were inspirational, and where the department leadership could only be described as mickey mouse. His final (as in "last straw") department head began her first faculty meeting by forcing the faculty to spend time tracing and coloring "hand turkeys" of the type drawn by first graders at Thanksgiving, to be displayed in the department headquarters for reasons neither ANDREW nor I will ever understand.

GRAHAM is another disgruntled former academic. When he married and moved, he had given up a position with lifetime tenure at one college to accept an untenured position at another. Despite a strong pedigree, good teaching evaluations, and research productivity, GRAHAM was denied tenure at his new college. His fatal mistake was to have spent some of his time at his new college publishing a pioneering article on how to better teach mathematics. His new colleagues believed that anyone who would spend time thinking about how to effectively teach mathematics must not be serious about doing "real" mathematics (proofs before pedagogy). Foolish; their loss, the country's gain. GRAHAM was ahead of his time: Not long thereafter, there was a national wave of interest on how to improve the teaching of mathematics and physics, a wave that has rolled on to include statistics.

Is there any return flow back from NSA to academia? I do not know enough to say. I do know that more than one Math Research person has gone back. My guess is that the Math Research staff at NSA generally feel that they already have the best of the academic world without some of the drawbacks. They can learn, they can teach, all their students are above average by construction and most already have Ph.D. degrees. They just get no spring breaks — but they are too old for that kind of thing anyway.

My suspicion is that the bigger threat of a reverse brain drain out of the NSA comes from companies like Google, Microsoft Research, Amazon, and some security-focused startups within the DC Beltway. These outside organizations would start to look attractive when the high cost of living in the Washington-Baltimore area starts to pinch, when technical people bump against incompetent nontechnical management, or when they themselves are forced by new personnel policies to accept management positions within the Agency. For a tech geek used to playing in a perfect geek sandbox, having to deal with budgets, schedules, recruiting, and personnel evaluation sounds like trading a heaven for a hell. At that point, the grass begins to look greener in a private sector R&D shop.

However, as the head of my own R&D shop, tiny though it is, I should warn them that the classic "go think about things" research operation is an endangered species in the business world. In my case, limited resources and staff and the need to keep our business running interfere with the relaxed thought that is required for technical creativity. In the case of formerly free-form, blue sky R&D shops like the GE Global Research Center, funding uncertainty, corporate reorganizations, and the need to charge hours (in 6 minute increments) to billable clients all crimp the style of the free-floating math geek. (OK, but the salary is a lot better outside, and private sector benefits are now competitive with government benefits, and nobody hates you for listening to their phone calls even though you are not ...)

Air Gap

There is another advantage to plying one's trade inside the NSA that appeals to many otherwise other-worldly academic types: Mission relevance.

Working at NSA, there is an immediacy to the sense of mission that is difficult to reproduce in academia. If nothing else, NSA works on problems of high national importance. There is a secret daily news summary published electronically within NSA called "NSA Daily". All one has to do is to skim these stories to understand the full urgency of the NSA mission.

I found that I could not make it through the day without reading "NSA Daily". It was my secret addiction. If I were feeling mentally sluggish in the morning, I could start the day with "NSA Daily" and get motivated. If I needed a break at lunch, I could read a few stories and get re-motivated. At the same time, I would often come away depressed and discouraged by the unrelenting flow of stories about attacks, threats, weaknesses, and dire possibilities. Friends and family would be surprised when I said that the work in the secret world could be depressing; that's what I meant. It is so easy to go merrily through the day on the outside and, even watching the nightly news, never understand the full dimensions of the threat matrix. Since I grew up during the height (depth?) of the Cold War and lived near a Strategic Air Command base that was a prime Soviet nuclear target, I have the notion of "threat" baked in. But nothing says "threat" like the "NSA Daily". No hype, just a catalog of grim facts.

This is not to say that academic work is not of high national importance. About one week into my sabbatical I had an epiphany. I realize that no matter how successful I might be with my Top Secret projects, all the years I had spent educating and grooming a high tech work force of statisticians and industrial engineers would ultimately do more to strengthen the country than whenever I could accomplish in a one year sabbatical.

(I have always found it particularly irksome that almost all portrayals of professors on TV and in movies show us as twisted, petty, salacious, mean, evil. One partial exception was "Indiana Jones". Most moviegoers who even remember that Indy was a professor remember the one scene in the first movie when a beautiful student winks at him. Experienced academics will tell you that the winking scene is just goofy. The real cool bit happens a little later, when Professor Indy is in the hallway talking to the government agents by a blackboard — and he pulls a... wait for it... pristine piece of chalk out of his... wait for it... tweed jacket! Classic! That bit comes and goes very quickly, but it's the best academic movie scene ever.)

Despite all the potential for the work of Math Research to bear on critical security threats, there is an obstacle to realizing that potential: the air gap between research and operations.

Math Research is a tiny sliver of the NSA. Most of the work is ground out by analysts ("analysis"), the people who feed the analysts their information ("dissemination"), and the people who manage the systems that collect that information ("collection"). Yet there is less contact between the researchers and the operational people than there should be. This inevitably means that researchers like me will be tempted to indulge our personal interests in cool things to think about, whether or not they relate to the urgent operational needs of the Agency.

During my sabbatical year, I had an intense desire to meet with and understand the thinking of analysts. However I was told that analyst time is precious, and during the entire year I met only once, for just one hour, with analysts. The analysts were a husband and wife pair working on a high priority problem. We did not solve their problem (they presented us with a desperate "Hail Mary" type idea that we could not see how to make feasible), and I did not gain much insight into their way of thinking. (What I did gain was an appreciation of just how hard a target ▮▮▮▮▮▮▮▮▮ had become.) This rare opportunity to be face-to-face with actual analysts owed much to the fact that GRAHAM was one of the small number of Math Research staff who worked hard to forge links with the operational people. Not long after our meeting, the husband suffered a stroke. The life of an analyst is not without constant pressure.

Perhaps because I was old enough to realize that organizational culture matters, or just because I was old enough to lack the mental stamina to work continuously on difficult technical projects, I kept returning to the problem of the air gap. I knew that the problems were urgent and that the research community had the potential to make great contributions to the operational community.

I understood that many in the research community thought of their mission as an abstract effort whose payoff might not come for years. Some also tended to eschew short-term research, possibly as undignified and unseemly. (Note that this would not be a widespread problem in the ranks. A key part of the recruitment process is to screen out math types who work for their own enjoyment and do not play well with others. But some get through.) This long-term perspective is especially characteristic of cryptologists. They were described to me as having a personality type that took a long view of things and were single-minded sometimes to the point of pathology.

I remembered being told about "cryppies" working the Russian problem during the Cold War. The Russians were and are world-class cryppies, so our folks would beat their heads against granite for years on end trying to break Russian codes. One can only admire that kind of persistence, but it created a management problem: Even cryppies need promotions, but they could go through years of hard and brilliant work without any measurable achievements. So the Agency adopted a clever accommodation in its promotion policies. It would transfer a deserving cryppy to some unsophisticated third-world target, whereupon said cryppy would slash through all kinds of slipshod codes, whereupon said cryppy would earn a promotion, whereupon said cryppy would be transferred back to the Russian problem for a few more frustrating but also mesmerizing years.

There are two sides to the air gap. I understood that many on the operational side were harried and thought they had no time for eggheads who had nothing practical to contribute to their problems. This is a situation ripe for proper management by the top echelon of the Research Directorate. Since I was at the bottom echelon, and even worse, an outsider-insider, I could do not do much to move the cultural needle. That did not stop me from thinking, however, and one idea that I actually voiced was to establish a sort of speed dating event, in which an operations person and a research person would sit down together for five minutes. The operations person would describe their problems, and the research person would try to think how research could help. Then a whistle would blow and the research person would move to the next station, as in real world speed dating (Not that I have any personal experience was speed dating, for the record.) Nothing came of my suggestion. Now that I think of it, a more promising alternative would have been setting up actual speed dating events, possibly leading to actual weddings between operations and research people. In that case, there would be some hope for mutual aid.

A few years after my sabbatical, while I was out of the DARKROOM and supporting NSA at IDA/CCS (see below), the air gap became even larger. The pressure to support the wars in Iraq and Afghanistan and other ventures in the Middle East lead to the expansion of certain ███████████████████████ groups in the ████ building. In response, Math Research was moved off of the Fort Meade campus to a suburban office park in Laurel, Maryland.

In the past, a researcher could take the NSA shuttle bus across campus over to the operations buildings; on a good day this trip might take 10 minutes. There is now a shuttle bus from the main NSA campus to Laurel, but this trip takes 30 minutes. In addition to the extra time and distance, now the researchers are sequestered in an isolated, leafy campus, closer to an old mill complex converted to a shopping mall than to the action back at the Fort. It is now much easier for researchers to talk to each other across parts of the Research Directorate, but harder to talk to the people who could rightly be described as their customers.

(The suburban location has a ████████████: There is ████████████████████████ One summer day there was a ████████ caused by ████████████, and the ████████████ had to ████████████ I didn't see any ████████ around the new DARKROOM. Poor KENNETH was already late to get home and relieve his wife, who was counting on his help with their very colicky new baby. Eventually we all ███████████████████████ but it made me wonder a little about ████████ issues. On the other hand, we had some serious guards between us and the outside world.)

Kill Chain

Few sabbatical opportunities involve death. This one did, in a way.

NSA is part of the Department of Defense. The DoD deals in death on a large scale. But NSA is also part of the Intelligence Community ("IC"), which does not normally deal in death (except for parts of the CIA). So spending time in a Mathematics Research group inside NSA can seem fairly far removed from dealing death. Day to day, what one sees are equations, graphs of equations, computer code, data, data and more data. One also sees technical courses, technical seminars, and technical papers.

Only twice did I hear anything in any way related to death. The first is classified. The other instance was a general announcement about the Memorial Wall that holds the growing list of NSA personnel killed in the line of duty. The CIA has a much better known wall of this type; I travelled from the ▮▮▮ building to the "Big 4" operations buildings (the ones you see in news stories about NSA and on the front cover of this book) to see a name added in a very sad ceremony.

NSA's Memorial Wall, listing the names of nearly 200 military and civilian cryptologists who made the ultimate sacrifice [https://www.nsa.gov/search/?q=NSA+Memorial+Wall]

While there is almost no whiff of death in the daily business of the NSA, it is obvious to the casual observer, or at least to anyone with a minimum of moral awareness, that the work we did was part of a kill chain. Knowing this required that I force myself, before starting the sabbatical tour of duty, to acknowledge this fact and affirm my acceptance. I thought of my father in combat in Germany and his discomfort with that memory, which involved, among other fraught moments, a recon mission that turned into an ambush that turned into a counterattack on a machine gun nest that turned into a Silver Star. I thought of the people I saw on TV who were forced to jump out of the burning World Trade Center towers. I thought of my wife, daughter and son and the people who considered them targets. I said yes.

The work done in Math Research is at the very distant end of the kill chain, but it is still in the chain. Some parts of the work are farther removed from a trigger pull than others, some closer. The first project I undertook was an attempt to develop an improved method of ███████████████████████████████████████ ███ At the time, the Iraq war was raging, and ███ was relatively close to the bloody end of the kill chain. I do not know whether my work was ever implemented and resulted in enemy killed in action (EKIA), and I will surely never know. Not knowing is not the same as not wondering.

I wrote earlier of the air gap between research and operations. There is another gap, between research and development. I prototyped my new ████████████████ algorithm in a computer language called R. (I taught myself R during the sabbatical, then came back to Rensselaer and forced my students to learn to use it, I believe much to their benefit. So from the university's perspective, the sabbatical investment was a double win: they had to pay only half my salary, and my students were better prepared.) Now R is a wonderful language for statistical programming, and I do all my company's prototyping in R as well, but it is not a wonderful language for production work. R is "interpreted" rather than "compiled", which is great for program debugging but too slow to put into operational systems. Any prototyping done in Math Research using R or Mathematica or certain other programming systems must usually be converted to C or some other fast language used in operational systems. As I well know from my own company, it is one thing for some person to cook up an algorithm to compute something. It is another to convert that analytical core into a useable software tool. Unfortunately, there is no large group of scientific programmers, systems programmers, and user-interface designers available to do the code conversion and then the further work needed to convert a good, fast piece of analytical code into a full software system usable by humans. As a result, I believe many promising research ideas sit idle for lack of programming help.

A lighter aspect of this project was my attempt to give it a cover name. Most, but not all, cover names are ███ and serve as a ███ that cannot otherwise be mentioned in unsecure settings. I don't believe the stories that cover names are ███, there are often ███ of cover names for ███ and perhaps the ███ are selected at random, but then the others are selected ███ Snowden's revelations ███ cover names, a few of which I expected to never see displayed in public in my lifetime, others of which just sounded rather goofy.

Let me make up an example. Let's say Project X is to listen to ███ and infer ███ (Death metal? Watch out ███ Viennese waltzes? Watch out ███ We might assign Project X the cover name ███ Then when Project Y is set up to ███ to see how much ███ it will get the cover name ███. Next up in sequence might be ███ and so on.

Back to my project. I was told I could give it a cover name of my own choosing, as long as it was unique. It turned out that there is a ███ cover names. I guess my concerns about killing people or spilling secrets or leaving secret documents out overnight was getting me down, and I devoted an unseemly amount of time to thinking about picking a cover name. Inspired by a certain TV show, I fell in love with ███ as a cover name. Off I went to the cover name ███ only to find that ███ had already been used ███ — how did they get away with that? Then everything else I tried was also ███. I finally ended with an innocuous choice: ███. Neither cool nor satisfying.

Another of my projects was located much farther from the sharp end of the kill chain. I was occupied with another project when KEVIN dropped by the DARKROOM to ask for help with something he was working on. KEVIN is a poster boy for NSA and the kind of public servant we can all be proud to support with our tax money. Formerly in the Research Directorate, he switched over to the operational side and served as a vital bridge between real world problems and the math research community. I did not know what he was really working on: He couched the problem in fairly general technical language. To do KEVIN a favor, and to enjoy his cool technical problem, I started working on his problem as a second task (I know myself well enough to appreciate that I work best if I have two or maybe three problems to think about at once. When I get stuck on one problem, I ricochet off to another. Things keep moving, so I can maintain the illusion of progress.) KEVIN's problem was too complex (for me at least) to solve analytically, so I used the engineering approximations I typically rely on when somebody needs a real answer real fast. Then I forgot about it – just another fun exercise that disappeared into the secret dark.

Back to the kill chain. Years later, I learned that my little throw-away project may have been the most important thing I contributed to the IC. KEVIN had incorporated my bit into a larger system and made it operational through the use of a ███████████████ (apparently Math Research does not usually have that option, and suffers as a result). When Edward Snowden exposed a number of NSA secrets, many high priority targets changed their behavior and disappeared from the view of our main system for tracking them (see, e.g., NSA Deputy Director Rick Ledgett's 2014 TED talk *https://icontherecord.tumblr.com/tagged/Richard-Ledgett*). It happened, though, that KEVIN's system provided a partial backup, and we were able to continue to track many of the targets. So, without knowing it, I helped limit the damage from Snowden's treachery. It is possible to imagine, therefore, that some of the ███████████████ were later visited by bullets or bombs, making me an indirect contributor to their deaths. Not a happy thing, but I am comfortable with my distant and unwitting contribution.

The Real People of NSA

For good reason, most of the people who work at NSA are invisible and want to stay that way. Those at the top are visible, on the news and as commentators. We would occasionally actually see DIRNSA (Director, NSA), often in a telecast from the Friedman Auditorium after a Big Event (e.g., the death of Bin Laden, the revelations of Snowden).

DIRNSA. The first DIRNSA I served under was Army general Keith Alexander. He seemed to sincerely care about his people. The first of the two times I was physically close to him was when he took the time to attend the retirement ceremony for ROSIE (see below). The second time, I almost barreled into DIRNSA in my rush to get to a men's room when he visited the new Research Directorate site in Laurel. He and his entourage came around one corner, me another. I'm sure if I had gotten any closer than I did, I would have found myself face down on the floor at the hands of his PPD (personal protective detail or "people putter-downers", take your pick). The resulting puddle of red and yellow body fluids would have been very embarrassing, yet surely also legendary.

ROSIE. ROSIE was the secretary in the office that handled the sabbatical program. She also worked on NSA's program of unclassified summer research grants to mathematics faculty. She always took excellent care of her two lost boys, myself and ANDREW. For instance, for some strange reason I was assigned a rank of GS15-step 10. That is the highest GS level before the Senior Executive Service and made me the civilian equivalent of an O6 or colonel. I discovered that the housing allowance for actual colonels at Ft. Meade was larger than the allowance that ANDREW and I were receiving. ADELE agreed that this seemed unfair, and ROSIE worked the system to increase our allowance. Before that, the allowance was not covering the rent on my apartment, and I think the situation was more critical for ANDREW, since he moved his wife and five of his six kids with him. (By the way, no one has ever seriously called me Colonel, and now that I'm not one (see below), no one ever will. But it was fun while it lasted.)

ROSIE herself is an interesting story. She spent a full career at NSA, and during that time she burrowed in very deeply. Some people know that the comedian Wanda Sykes had at one point worked in a clerical capacity at NSA, and she may be the most visible black alumna of the Agency. But ROSIE stayed in a long time, and I noticed that she was a key member of a large circle of black women who probably ran the Agency behind the scenes. ROSIE was the one who, on Day 1, retrieved me from the Visitors' Center and guided me through the intake process (escort through the maze, photograph, security briefing, swearing in, assignment to office, payroll paperwork, etc. etc.). Literally everyplace she went, with me in tow, she would see sisters from her circle. If the sergeants really run the Army, and the chiefs really run the Navy, then the Sisters may really run the NSA.

Most of the other black faces at NSA belonged to the many uniformed troops who seem to comprise the bulk of the work force at Fort Meade, and color seemed to have no significance to those in uniform. I saw few people of color in Math Research or other of the technical hotspots within NSA. This reflects the complexion of the entire STEM (Science, Technology, Engineering, Math) pipeline, though that is changing by demographic necessity alone. At Rensselaer, I have seen a gradual increase in the number of minority students in my statistics classes. I also saw one prominent payoff from the special counseling that we gave to our minority students. They would come early to class, sit in the front rows, not fall asleep, and even ask questions. That kind of "leaning in" paid off for them. (It also did wonders for my morale; few things are more discouraging than trying to teach the dead. Having at least a few students with a pulse makes it so much better.)

Math Research sits at the very end of the STEM pipeline, where the entry level credential is most often a Ph.D. degree, usually in mathematics. (Math Research could benefit from more engineering Ph.D.'s, though perhaps better ones that me.) I have no doubt that NSA is trying to work on its diversity problem, but I know that recent budget cuts have reduced the amount of college recruiting that the Agency can do, and negative reactions to Snowden's revelations have no doubt also hurt the effort.

<u>The troops</u>. In general, my impression is that the troops fell on both extremes of the spectrum of passion for the work. Some were there just because they were assigned; they knew they'd be moving on and did their jobs, period.

Some felt tricked. I remember hearing from a young woman who was in charge of about thirty sailors. These sailors had enlisted to "work with computers", by which they meant working with their hands on hardware: soldering, connecting, etc. Instead, their mission was to sit at a web browser and, without sufficient direction, "look for badness on the internet". (My team's project was to ███████ ███████████████████████████████████████ so they wouldn't ███████████████████████████). Her sailors had little motivation and did not work hard. I asked why she didn't get their chief to straighten them out. She said the chief was a problem too. The chief needed good performance reports for his sailors for the sake of his own career, and he himself didn't show much enthusiasm for his NSA job. So he would invent "real Navy" side jobs for this sailors (I never understood what those made-up jobs might be on dry land, but you don't get to be a chief without being able to solve problems) and give them excellent fitness reports focused on those. Then everybody could move on to "real Navy" jobs when they rotated away from Ft. Meade.

At the opposite end on the passion scale, other troops felt a deep personal connection to what was happening in Iraq and Afghanistan, either because they had been there, they were going to deploy there, their buddies were already there, or all of the above. Earlier, I mentioned the multitude of seminars in Math Research. There were also Agency-wide seminars, which I tried to attend whenever there was a hope of expanding my sense of the big picture. One I will never forget was given by a Marine major on the subject of defeating the IED's that were the major source of our casualties in the two wars. Never have I seen such controlled but intense passion in any seminar on any subject in any venue. It was easy to imagine that he had lost men and expected to do so again. He was pleading for help, welcoming any wild and crazy idea that might have a prayer of being effective. This seminar helped me see not just one of the areas where Math Research could make a large contribution but also to recognize that NSA's support for the war fighter is part of its mission to thwart the other guy's kill chain and save lives. I was frustrated by not being able to offer any ideas to that Marine.

<u>Invisible people</u>. What about the rest of my colleagues? By design, most NSA employees are anonymous or try to be (though I was informed in my first week that the Russians and Chinese would already have my name). I was told there is a standard protocol to follow when somebody asks "Where do you work?" The first answer is "For the government". In the DC area, people are assumed to recognize that code and drop it. If they ask again, we could answer "For the Department of Defense." If they persisted, we could surrender and say "For the NSA." But that was it.

One could imagine that certain bad people would like to know more. I assume that an NSA phone book, listing all who worked there, would fetch a high price on the traitor market. It would be no surprise if sometimes foreign intelligence services make an attempt.

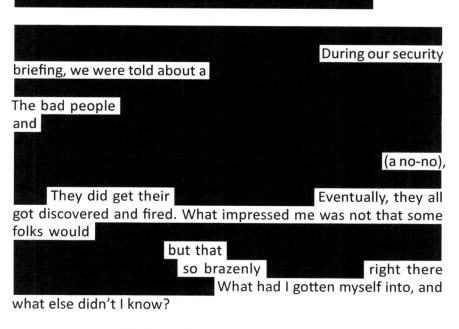

During our security briefing, we were told about a The bad people and (a no-no), They did get their Eventually, they all got discovered and fired. What impressed me was not that some folks would but that so brazenly right there What had I gotten myself into, and what else didn't I know?

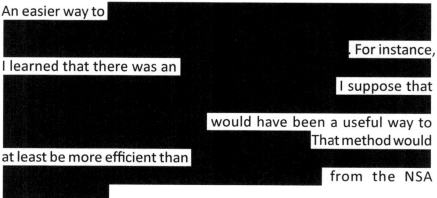

An easier way to . For instance, I learned that there was an I suppose that would have been a useful way to That method would at least be more efficient than from the NSA

It's easy to understand why other countries' spies would treasure a list of NSA employees. Chris Inglis, former Deputy Director of NSA (DDIRNSA) and therefore the top civilian, told Steve Inskeep in an NPR interview: "… if you want to really know what the core of NSA is, it's its brain trust. It's its people." [http://www.npr.org/2014/01/10/261282601/transcript-nsa-deputy-director-john-inglis]. But it is difficult for America to appreciate NSA professionals when they mean to be invisible.

Some resent that it is difficult to know the "brain trust". But with a little work and a few stolen classified documents, it appear possible to overcome the secrecy. Peter Maas wrote "… one of the asymmetric oddities of the NSA is that the agency permits itself to know whatever it wants to know about any of us [untrue but commonly assumed – TRW], yet does everything it can to prevent us from knowing anything about the men and women who surveil us, aside from a handful of senior officials who function as the agency's public face." [https://theintercept.com/2015/08/11/surveillance-philosopher-nsa/] Nevertheless, Maas then went on to research, with the help of Snowden-released materials and open-source information, one NSA person, revealing many details of his or her life and family (though Mass stopped short of naming the person). I came to admire the NSA person based on his secret blog, but I do not admire the effort to (almost) out him.

John Dean (yes, that John Dean), a famous close observer of bad behavior in government, asked: "How do the people who work at NSA who were turning that huge electronic apparatus of surveillance on their neighbors and their friends, where's their conscience?" [http://www.greendel.org/2009/02/12/do-we-all-have-the-capacity-for-inhuman-cruelty/]

In contrast, Representative "Dutch" Ruppersberger (D-MD), whose district includes Ft. Meade (and whose opinions are therefore discounted by those who vilify the Agency), expressed this positive opinion: "… these people who work at NSA are hard-working people who follow the law. In fact, we have lost 20 members of the people working for NSA in Iraq and Afghanistan attempting to get information to help the troops." [http://abcnews.go.com/ThisWeek/

week-transcript-gen-martin-dempsey-reps-ruppersberger-king/ story?id=19864913]

Like Dutch, I too have this bias, though mine comes from observing the lower rather than higher echelons of NSA. Without naming names, let me provide a few vignettes and capsule descriptions to fill in the picture of what these shadowy people are like.

<u>GRAHAM.</u> We've already met GRAHAM. GRAHAM is focused on providing practical solutions to real-world problems. In this, he shows that he has overcome the "disadvantage" of his education. His doctorate is in a distinctly impractical corner of abstract math, and he was supervised by an advisor who was observed to not only start a fire in his own office but to also actually walk smack into a tree whilst thinking deep thoughts. Going back even further, GRAHAM grew up in the "wrong" part of Chicago for a white kid, but his music abilities got him in with the black kids. For recreation now, he plays his instrument a little and reads philosophy. Besides showing me the ropes within the DARKROOM and collaborating with me on several top secret reports, GRAHAM showed me the value of persistence. He is a bulldog when working a problem, even ones that seem beyond solution. When I started my sabbatical, there was a large sign in the lobby of the R&E building that read "We never back down. Never have, never will." GRAHAM lives that sign.

<u>HENRY.</u> We've also already met my former next-desk neighbor HENRY. He is a great source of stories about how the Agency used to be. HENRY recalled as a kid driving right up to the NSA building where his dad worked and having a picnic lunch on the lawn. Now that the truck bomb has been invented, of course, it is not possible to do anything so informal. Getting near the Agency is difficult. Besides the fences and the strategically placed boulders, NSA maintains a highly lethal police and guard force. HENRY showed much more intellectual stamina than I. As I withdraw from my technical life during my Social Security/Medicare days, HENRY continues to eagerly dive into totally new technical areas. At this point, he has retired, but NSA has a policy that allows retired experts to continue

to contribute part time, so HENRY just rolls on. Those who doubt the dedication of federal employees have not met people like HENRY.

COLIN. We have also met COLIN, my Technical Director (TD). Despite his near pathological but always clever punning, COLIN set the template for the technical manager as player/coach. He continues to win prizes and awards for this technical work, but he also watched over his portion of Math Research with an eye to making sure that all his people are happily and productively engaged. No longer a TD, he is finishing his long career at the Agency with continued technical and leadership contributions.

LINCOLN. In some ways, LINCOLN comes closest to fitting the image of the uber-geek. Nobody takes longer to complete an analysis, yet nobody penetrates deeper into the dark heart of a statistical problem. I regarded LINCOLN as a living encyclopedia of deep knowledge of probability and statistics. The contrast in our work styles could not be starker. Trained as an engineer and shaped by drop-dead deadlines as a software entrepreneur, I aimed to provide a "good enough" answer quickly enough to influence a decision. (Truth be told, "good enough" was probably the best I could do anyway.) LINCOLN, on the other hand, would often start out by saying "Let's do the right thing here" and then methodically grind out a minutely and comprehensively wonderful analysis, often supported with *Mathematica* code. Conversations with LINCOLN resemble his analyses. I remember once standing with my hand on the DARKROOM doorknob for over 30 minutes while LINCOLN, not taking the hint, spun out an admittedly fascinating story about sports, the other subject about which his knowledge is encyclopedic.

LLOYD. LLOYD recently replaced COLIN as one of the TD's in Math Research. Like several others, LLOYD is an expert in statistics who retooled himself after training as a computer scientist. ███████████████████ he was forced early to find his way to work/life balance, ███████ (like my own granddaughters) are an inherently unbalancing life force. LLOYD's leadership style will feel different than COLIN's. I speculate that will look a bit more businesslike, and I expect it will lean, at least marginally, more toward solving problems of immediate operational interest. LLOYD's initiative in establishing the Thursday Statistical Scientists' Meeting bodes well for his time as a Math Research TD.

Early in my sabbatical, I met LLOYD in a sandwich place in Princeton, NJ. I was attending my first classified technical conference. I tend to regard meal time as reading time, so I supplemented my turkey sandwich with a few pages from my ever-present copy of *The Portable Medieval Reader* (James Bruce Ross and Mary Martin McLaughlin, editors), a compendium of translated works from the Medieval period. (One of my favorite excerpts: A piece by Roger Bacon, the 13th century's "Doctor Mirabilis", making the case for objective experiments in scientific research, but then concluding with his observations on the flying dragons he was certain must exist in North Africa.) LLOYD noticed the book and confessed that he too was devoted to it. It was comforting to discover another oddball on the inside.

This particular technical conferences rotates through venues. That year, it was held at the Institute for Defense Analyses/Center for Communications Research in Princeton, NJ. IDA/CCR-P is the oldest of the three IDA outposts that support NSA (the others are in Bowie, MD and La Jolla, CA; much more on Bowie below).

CCR-P is now located in a semi-rural part of Princeton, but my first incidental encounter with it was as a Princeton undergraduate, when it was adjacent to the Princeton University campus, midway between the engineering quadrangle and my Princeton eating club, Cloister Inn. It sat behind a high brick wall. Unaware, I walked by it countless times between lunch and lab without giving it a thought.

I only became aware of it when a group of students in the Princeton chapter of Students for a Democratic Society (SDS) protested its existence, its oversight by the university, and its presumed (but mistaken) support for the Vietnam War effort. During the protest, one of the CCR staff members took a swing at a student demonstrator, who ducked, resulting in the punch cold-cocking the man standing behind him, Dr. Neil Rudenstein. Rudenstein was then a Princeton administrator and later president of Harvard. Perhaps I should have realized right then that my future self would get entangled in an IDA adventure and begun practicing my "IDA combat bob and weave". (The SDS chapter later won a challenge touch football game against an ROTC team. One of the SDS members in my Class of 1969, Steve Weed, became a footnote to history years later when the Symbionese Liberation Army cold-cocked him and abducted his girlfriend, one Patty Hearst, a.k.a. "Tanya".)

These historical tidbits do not exhaust the back story of IDA/CCR-P. I learned at a banquet concluding the research conference that IDA/CCR-P later moved away from campus to a building elsewhere in Princeton that was partly owned by, wait for it… the Communist Party of the United States. (For more on IDA/CCR and the student protests there during the Vietnam war, see Dr. Lee Neuwirth's book, **Nothing Personal: The Vietnam War in Princeton, 1965-1975**.)

The year I attended, the conference was in the present, third, location of IDA/CCR-P, safely away from all communists and proto-communists. The ▮▮▮▮▮▮▮▮ conference was an eye-opener for me, but also somewhat awkward. Still new and unsure of security protocols, I did not know how or where to engage other attendees in substantive conversations, and I always felt odd-man-out amidst people who all seemed to know each other well. Social awkwardness aside, I got to sample the full and overwhelming range of research that was ongoing across the Intelligence Community. I also got calibrated about the high technical level expected of the work I would do. The lively debate that followed most presentations was exciting and contrasted with the disappointing passivity common at nerd gatherings on the outside.

CHRISTINE. One summer, CHRISTINE served as the leader of my ▮▮ In that role, she modeled the ideal team captain. She would make the rounds of our offices every week, asking whether we had all we needed to accomplish our tasks. In fact, she devoted so much time to helping everybody else that she barely had time to create her own final report. Luckily for me, part of her effort was helping me understand the problem I was working on, and we ended up co-authoring another of those Top Secret reports that either disappeared into obscurity or actually helped solve a problem, in this case helping those aimless and disgruntled sailors ▓▓▓▓▓▓▓▓▓▓▓▓▓▓▓▓▓▓▓▓▓▓▓▓▓▓▓▓ CHRISTINE has a sly, mischievous sense of humor, which helps the mood in the new, suburban version of the DARKROOM. At one point she photo-shopped a very funny poster based on one of those famous pictures of a bare-chested Vladimir Putin. Perhaps the funniest part of the episode was her sudden concern that she might get found out and treated like a bad little schoolgirl during a surprise visit by some bigshot (she escaped, one hopes to strike again).

CHRISTINE is half of one of those NSA marriages. Her husband ▓▓ ▓▓ Once, CHRISTINE and her husband joined HENRY and his wife at my apartment for (what they didn't know would be) a birthday party for me, thrown by my wife in one of her infrequent visits (My motto: "Even prisoners get conjugal visits.") As a former Distinguished Visiting Professor at the FAA, I continued to indulge my penchant for listening to air traffic control conversations (which must qualify as my only actual hobby, if it even qualifies as a hobby). In the kitchen, I had an aircraft scanner running; the sound was turned off, but the LED lights were flashing away as pilot/controller conversations came and went. Wandering into the kitchen, HENRY discovered the scanner running. I later found out that, always alert to the need for security (and maybe not knowing me for long enough to trust me), HENRY got worried that I was secretly recording all the party talk in the other room. This episode is a good reminder that NSA business is serious, and that there is an affirmative duty to vigilance in the protection of secrecy. It also reminds me why I was visited by nightly nightmares about security blunders during my first weeks at the Agency.

<u>LUKE.</u> When I began my sabbatical, LUKE was the head of all of the Research Directorate. I gather from conversations that my colleagues had mixed feelings about the various worthies who had previously headed Research Directorate over the years. In LUKE's case, there seems to be a consensus that he was a wonderful leader (he has since retired). Maybe LUKE achieved his stature by imprinting all those interns the same way Konrad Lorenz imprinted his ducklings. LUKE taught a famous and much-loved introductory course, designed to convert hopelessly theoretical mathematicians and computer scientists (Is there a difference these days?) into effective practitioners. Well, there must have been more to it than that, because the old hands also held LUKE in high regard.

One of my proudest moments was when I was summoned to LUKE's suite and presented with his "challenge coin". At the time, I did not know the significance of this exchange. Wikipedia informs us that:

> "A challenge coin is a small coin or medallion (usually military), bearing an organization's insignia or emblem and carried by the organization's members ...challenge coins are normally presented by unit commanders in recognition of special achievement by a member of the unit."

They are called "challenge" coins because one party will display his hard-earned coin to another and challenge him to show his own. If the challengee cannot produce a coin, he is obliged to buy a drink of the challenger's choice. If the challengee does produce a coin, then the challenger is obliged to buy a drink of the challengee's choice.

That I received LUKE's coin for something I did during my sabbatical remains a source of vague pride. Pride, because I must have done something worthwhile. Vague, because I have no idea what good thing I did. Perhaps the secrecy surrounding awards is sometimes so extreme that even the awardee knows not what he has accomplished. On the other hand, one of my colleagues deflated me a little by opining that the Big Guns were so surprised that any sabbatical visitor did anything of even modest value that they awarded the coin in a reflexive act of astonishment. However well or accidentally earned, I have not followed the strict military tradition of having my coin available at all times, including in shower or latrine. Maybe I should start carrying it around when I try to sell software to the armed services. They seem to have no budget for improving their logistical efficiency, but they sure do a lot of drinking.

Challenge coin presented by the Chief of Research Directorate

Challenge coin after obfuscation per NSA mini-course on "Denial and Deception"

MATTHEW. When I began my sabbatical, MATTHEW was the head of Math Research, which probably means that he shares some of the blame for OK'ing my selection. Though slightly less a mystery to me than LUKE, he played a major role in my ability to continue to contribute to the IC after my sabbatical ended (see below). MATTHEW was one of those quiet superstars who had major technical chops but also level-headed management skills. In one important example of bridging of the air gap, MATTHEW left the Research Directorate for a critical operational assignment running a major component of the ███████████████████████████████████ side of the house.

NELSON. NELSON and I sat through the same electrical engineering courses at Princeton, then we parted ways for many decades until crossing paths again at NSA. NELSON devoted a full career to the Agency, and even after retirement continued to work as a consultant. One of his more historic accomplishments was setting up the first "hot line" between the US and USSR.

At the end of his career, NELSON moved on from the reassuringly tangible world of communications systems and red phones to wrestle with a big fuzzy problem: establishing what he called a "science of security". Part of this work involved strengthening the American academic infrastructure in computer science, so that there would be a pipeline of qualified people with a more practical and security-minded perspective on that field. I was of minor use to him in that task, because I could simulate the reaction of a professor to some of his proposals. (Start with "What's in it for me?" Maybe he didn't really need me after all...) The hard part of his mission was thinking about what a "science" is, what "security" is, and what it takes to make a new academic field. NELSON worked the whole world in that assignment, tapping Oxbridge professors in the UK, meteoric superwoman Jeannette Wing at NSF (later VP of Microsoft Research), and, in a pinch, even me. I admire his intellectual courage, as I would have run screaming from such an assignment.

NELSON often gives off the vibe of the prototypical grumpy old man. However, he was a godsend to me in two ways. First, he helped me understand the inscrutable ways and mores of the NSA. Second, he fed me! NELSON is a native Baltimorean, so he knows all the best places to eat. He instructed me in the cosmic importance of Old Bay spice. He taught me and my wife how to eat crab. (She loves the process and is good at it. I think it hurts too much when all that Old Bay gets in the cuts on my fingers from those sharp-edged shells. Give me a lobster instead – there's a bigger, sweeter payoff for less finicky work.) I have good memories of the times he invited me to dinner at his suburban house, breaking the loneliness of my time in the IC. The man knows international hot lines, and the man knows how to do steak.

NELSON's surface gruffness helped him blend in to his surroundings in the Information Assurance Directorate (IAD). NSA has twin and contradictory missions: to steal the secrets of potential adversaries and to protect our own. NELSON's part of NSA, IAD, works the side that protects military and civilian computer infrastructure. The NSA web site describes the (sometimes contradictory) missions of the Agency this way:

> *Mission Statement*
>
> The National Security Agency/Central Security Service (NSA/CSS) leads the U.S. Government in cryptology that encompasses both Signals Intelligence (SIGINT) and Information Assurance (IA) products and services, and enables Computer Network Operations (CNO) in order to gain a decision advantage for the Nation and our allies under all circumstances. (*https://www.nsa.gov/about/mission-strategy/*)

Most of the time, there is no contradiction, because the secrets we steal belong to others and the secrets we protect are our own. However, sometimes the Agency finds itself in a predicament. If we discover a vulnerability in, say, a Microsoft or Apple product, do we (a) keep it a secret (a "zero day") and exploit it or (b) notify the vendor and protect military systems and/or American companies from hacking by foreign entities? This is referred to inside as the "equities problem". In difficult cases, a "vulnerability equities process" convenes meet to make the call. [See *Electronic Frontier Foundation v National Security Agency, Office of the Director of National Intelligence*, Case 3:14-cv-03010-RS, United States District Court, Northern District of California, San Francisco Division, Exhibit C; available at *https://www.eff.org/files/2016/01/18/37-4_richberg_declaration_ocr.pdf*]. (As I write this, the equities problem has exploded into public view in the conflict between the FBI and Apple about iPhone encryption.)

What I discovered, to my disappointment, is that the people NELSON introduced me to in IAD, unlike NELSON himself, seemed to be jaded, disgruntled, and defeated. The defeated part is somewhat understandable, since in cyberwar the defense always seems to take a beating. But I was disappointed to see, for the first time, people at the Agency with such visibly low morale. They seemed to think that their bosses were incompetent, and maybe they were right. They seemed to have devolved down to clock-watching, days-to-retirement-counting versions of the popular cartoon of government workers. I expect that these guys were actually doing a decent job, but there was a clear contrast with the mood and morale of the people in Mathematics Research.

<u>FAITH.</u> Not all the people in NSA have exotic jobs. There are a large number of indispensable support people (see ADELE and ROSIE above). One such is FAITH. FAITH was the chief librarian for Math Research. Thanks to FAITH, it was hard for me to find a technical journal too obscure or exotic for inclusion in our library. When Math Research lived on the Ft. Meade campus, the library was the size of a coatroom. When Math Research moved to its remote suburban office park, one of the few positives was that there was decent library space, and even reading tables. FAITH's job performance was excellent, but she brought something else to Math Research: a bit of pizzaz. Most people at NSA dress in a way that ranges from uninteresting to grungy. FAITH, though, has a wardrobe of infinite variety and style. I thought of her as a kind of moving floral display, adding some color where color was needed. FAITH, like so many others, was half of an Agency couple. But while FAITH was the quiet but stylish librarian, her husband was the █████████████

<u>The Ninjas.</u> NSA has an extensive security force. Some are ███, but many are ███. These were referred to as ███████████████

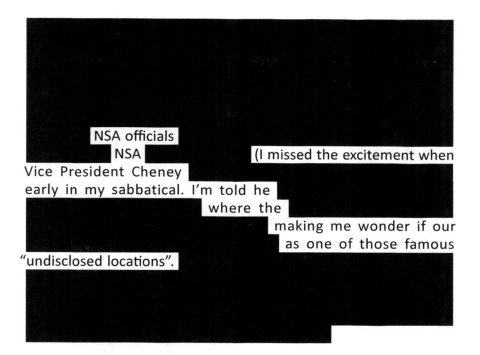

NSA officials
NSA   (I missed the excitement when
Vice President Cheney
early in my sabbatical. I'm told he
where the
making me wonder if our
as one of those famous
"undisclosed locations".

<u>The Five Eyes.</u> I had intermittent contact with people from four of the "Five Eyes countries": US, UK, Canada, Australia, and New Zealand. I don't recall crossing paths with a Kiwi, but British visitors were not uncommon, and a few Brits, Canadians and Aussies collaborated in projects I worked on. Without exception, the boys and girls from GCHQ, England's version of NSA, were impressively brainy, though usually jet lagged when I met them. Likewise, the few Canadians we worked with were highly effective. The one exception was attached to our team in SANDBOX (see below).

Working with Five Eyes partners could be fun. I remember a moment during a dinner in Annapolis. I said that it seemed from everything I'd read that every plant and animal in Australia was poisonous – even the dirt was probably deadly. A nearby Canadian said that about all they had to worry about in Ottawa was poison ivy. The Aussie looked concerned and asked how long it would take for poison ivy to kill a man. Laughter (almost) all around.

GRACE. ▮▮▮ I was launched: My name was on my first classified document. I was a little less outside, a little more inside.

GRACE ▮▮▮ (My Dad was a machinist and trumpet player, but for most of my youth in he was in VA hospitals for a service-connected disability.)

Others. Math Research, indeed the entire Research Directorate, is full of interesting characters. There was NICHOLAS, who is hands-down the best question-asker ever to attend a technical talk on any and every subject, anywhere. I've always considered it to be a great insult to the speaker if a seminar ends with no questions. NICHOLAS insured that Math Research is an insult-free zone. There was PETER, who wrote a textbook for an internal course in probability and statistics that has more personality and sparkle than any I've read on the outside. Too bad it will never see the light of day. There was PHILIP, now gone on with LUKE to the Place-Where-All-Go-After-Leaving-the-Agency-If-They-Don't-Want-the-Big-Bucks, i.e., the ▮▮▮▮▮▮▮▮▮▮▮▮▮▮▮▮▮▮▮▮▮▮▮▮▮. PHILIP invented a software program for visualizing terrorists' social networks. That and many other inventions made him the poster boy for government intrapreneurship. (My company is working to convert PHILIP's software to analysis of relationships within spare parts inventories, which we call the "social network of wing nuts". It is not widely known that the Agency licenses some of its software and other inventions to qualifying American companies. See the Technology Transfer Program's web page at *https://www.nsa.gov/research/tech_transfer/technologies/index.shtml*.)

Those not present. One person I never did encounter was the classic cartoonish Ph.D. mathematician wandering around in circles wearing pajamas and a bathrobe and talking to himself. (The one guy I saw doing this skipped the bathrobe bit but not the talking in circles bit.) Nor did I see any evidence of the preserved alien bodies supposedly moved to Ft. Meade after their journey from Roswell to Area 51 to Wright-Patterson Air Force Base. I can only speculate that they were moved to the CIA before Cheney came for his visit to Ft. Meade. (I have been to Wright-Pat twice trying to interest the Air Force in my software. I detected no alien traces, but I did see another rare sight there: A female one-star.)

SANDBOX: Something to Do

When I was scrambling around for something to work on early in my sabbatical, I was advised to join ████████, which I will call SANDBOX. A SANDBOX is a kind of slow-motion tiger team, assembled from various corners of the Agency to work over a period of six or more months on a single important technical problem. That turned out to be good advice, even accounting for the occasionally trying presence of our ████████, whom I'll call PRESTON.

We shared the rather tired SANDBOX space with another SANDBOX team, which at first seemed more interesting because it had at least one ████████ person in it. (To this day, she was the only ████████ person — that I know of — who has crossed my path. I'd been told that the names used by ████████ people when dealing with us were never their real names, but I never got to know even her fake name. All I knew was that she was Chinese-American and just as nerdy as the rest of us, so she must have come from their technology directorate.)

There was a little cross-talk between the various SANDBOX teams, and one of their members, RAYMOND, eventually spent at least as much time in our space as he did in his own. This mattered, because the combination of RAYMOND and PRESTON amounted to the single biggest noise source I've encountered in the IC. RAYMOND had come to us after joining and then very quickly (and very wisely) leaving one of the main sources of the subprime mortgage loan scandal that brought on the Great Recession. RAYMOND was brilliant and, like many a newly-minted Ph.D., not shy about letting it be known. He may have been, and may still be, the fastest-talking person ever hired by the Agency. His saving graces were (a) he is a nice guy and (b) he had some really insightful and stimulating ideas to share with us in SANDBOX. But since I had a goal in mind and felt some time pressure, RAYMOND was more often a distraction. (I only have so much bandwidth.)

But there was no bigger distraction than PRESTON. Perhaps during his earlier service in the much undermanned and overworked ▓▓▓▓▓▓▓▓▓▓▓▓▓▓▓▓, he developed the habit of talking too much as a way to make the work seem less tiresome. Perhaps that's just the way of people from ▓▓▓▓▓▓▓▓▓▓▓▓▓▓, of whom I still only know one. In any case, our SANDBOX project coincided with the presidential election campaign pitting Barack Obama against John McCain.

Political discussion is a no-no at the Agency: It is inappropriate, and there is too much work to do anyway. Apparently PRESTON thought the rules did not apply to ▓▓▓▓▓▓▓▓▓▓. Not a day went by without PRESTON stirring the pot, giving his personal opinion about all the issues in the campaign. I began to wonder whether the ▓▓▓▓▓▓▓▓▓▓ had loaned PRESTON to us just to give themselves a little peace and quiet. I also began to sit at workstations that were on the other side of the large pillars in the room, so that he could not see me and try to suck me into his discussions. Even if I had agreed with his attacks on Obama, just the fact that he was entangling somebody else on the team in his diatribes pissed me off. Our SANDBOX leader was way too lenient in letting this go on.

In the end, though, PRESTON was a lifesaver for me. Since I was an outsider who did not understand or in some cases have access to the many data sources available in the Agency, I was at a disadvantage. Further, since I had not used UNIX for many years and most Agency work was done using one or another flavor of UNIX, I had limitations in manipulating the data that I could access. I am pretty good at analyzing data, but heavily dependent in all my work in the IC on someone who can feed me data to analyze. PRESTON fed me data, so I was able to produce several reports that felt like contributions. Besides his own useful contributions to our problem, PRESTON saved me from being next to useless. (What's the human analog of a duck out of water? A statistician out of data.)

One of my reports spilled outside the Agency. I had invented a new statistic for characterizing computer files. (We were getting many ▮▮▮) My statistic was strange enough to be interesting, so COLIN and LLOYD in Math Research sanitized its origin and unveiled it to the outside world as an NSASAG problem.

NSASAG stands for "NSA Statistical Advisory Group". It is comprised of statisticians with national-level reputations who volunteer to help the Agency during summers by responding to difficult statistical questions. The Agency removes the classified context of the original statistical problem and asks for help. Many of them are on the faculties of leading statistics departments at ▮▮, though I believe some individuals dropped out of NSASAG in response to the Snowden affair.

Lacking the ability to do a classical mathematical analysis of my statistic's properties, I relied on some engineering approximations to understand its behavior. I must confess that I was secretly pleased to see that the NSASAG experts were not able to push the analysis any deeper than I had. As usual, I have no idea what became of the work of SANDBOX in general or of my little statistic in particular. There is a decent chance, though, that this work too became part of the quiet end of another kill chain.

I have two other strong memories of SANDBOX. One is my surprise at learning that whenever I wrote a classified report, I had to do my own classification. Furthermore and sometimes to my annoyance, I would have to make many such decisions, since I had to "portion mark" everything I wrote. That is, every paragraph, every figure, every figure caption, every reference, every entry in the table of contents, all components of a report had to be classified. I did not feel comfortable deciding between "unclassified" and "classified". I was told that sometimes the technology in question was not what was classified; instead, it was the very interest of the Agency in the topic that was classified. I also had difficulty distinguishing between SECRET and TOP SECRET information. The former is defined, vaguely, as information that would cause "grave damage" to the nation's interests if publicly revealed; the latter would cause "exceptionally grave damage". There is a lower category, "Confidential", which almost never figured in our work. There is a catch-all, "For Official Use Only" (FOUO) that got slapped on a lot of paragraphs, probably having in most cases only the effect of impeding legitimate citizen access by frustrating Freedom of Information Act (FOIA) requests. In addition, there are various compartments, which further restrict access. So TOP SECRET is not as secret at TOP SECRET NOFORN, which forbids sharing with foreign entities.

I was very uncomfortable making these classification distinctions, so I asked the woman who was the long-serving Godmother of All SANDBOXes whether a particular something was SECRET, TOP SECRET, or something else. The same question eventually got put to another person, whom I never saw, who was the Godmother of All Classifications. The Godmother of SANDBOXes and the Godmother of Classifications disagreed. Eventually, I got comfortable with being uncomfortable about making these distinctions. Later, working in SCAMPs, I was told that our reports would be ██ (As I write this, the subject of ambiguity in classification is a part of the 2016 presidential campaign. The CIA and State Department are in a public disagreement about the proper level of classification of some of former Secretary of State Hillary Clinton's emails.)

The other memory that lingers from SANDBOX was spatial. The SANDBOX space was on a high floor of the Old Headquarters Building. When you see pictures of the NSAW campus, the eye is drawn to the two big shiny buildings with the reflective windows (Ops 2a and Ops 2b), which seem massive and mysterious and vaguely threatening. Most people do not notice the older, dumpier Old Headquarters Building. But being in there, and circulating amidst the other "Big Four" buildings, put me physically on the other side of the air gap. It gave me a greater sense of the size and pace of the Agency. It made me feel less academic, since usually what I did as a professor and what I did as a Math Research person seemed very similar.

Being in the SANDBOX space also gave me a visual reference. Nowadays, if I find myself driving on Maryland Route 32, I can look up and see the very window that I used to peer out of when taking a break. In turn, that makes me remember that my view from the window took in the Capital Guardian Youth Challenge Academy located on the opposite side of Route 32. This appeared to be a prison for young offenders. There they were behind barbed wire, trying to make sense of their turbulent lives. There I was across the highway, behind my own barbed wire (more and better wire, more and better guards), trying to make sense of the turbulent world in which America operates under threat every day.

Naughty or Nice: The Dawn of ACE

(This chapter is a bit different from the others. It focuses on the subject of performance evaluation for knowledge workers, both in the IC and in universities. If you agree with me that this is an interesting and important topic, read on. If not, feel free to skip ahead.)

One of the benefits of being an outsider on the inside is that I was able to sidestep some bits of the Agency bureaucracy. One seismic shift that rattled my colleagues during my sabbatical visit, while leaving me unscathed, was the imposition of the personnel evaluation system called the "Annual Contribution Evaluation", ACE.

Every organization needs some sort of personnel evaluation process. The one at CCS (see below) appears to be blissfully simple, cheap, meaningful and fluid. It can afford to be, as CCS is small and its horizon is long. (Whether CCS's more relaxed process would help or hurt in an employment discrimination lawsuit I have no idea.) I don't know anything about the system that preceded ACE at NSA, but presumably it was of a size and complexity commensurate with the size and complexity of the Agency, i.e., pretty ugly. But the old system must not have been quite as ugly as ACE: Oh, the groans and moans when ACE was announced!

ACE is sufficiently elaborate to generate a good-sized flow of paperwork. Three people are involved: the Employee, a Rater, and a

Reviewer. The Employee must submit an annual performance plan with specific objectives. Besides the annual review, there is a Mid-Cycle Review. The first pages of the annual and mid-cycle reviews are shown below.

Employees are rated on both their self-generated "objectives" and on generic "performance elements" that are not employee-specific. There can be three to six objectives (probably less grandiose than "I will invent ███████████████████████", "I will destroy Daesh", "I will find American suppliers for the NSA Gift Shop"). All employees are evaluated against four elements: "Accountability for Results", "Communication", "Critical Thinking", and "Engagement and Collaboration". There are two more elements, depending on whether the employee is a supervisor. Non-supervisory personnel are rated on "Personal Leadership and Integrity" (notably including "the courage and conviction to express their professional views") and "Technical Expertise". Supervisors are rated on "Leadership and Integrity" and "Management Proficiency". The bottom line of each ACE is a number. The scoring system and calculations are shown below. Ratings range from 1 to 5 and are meant to be centered around 3.0.

I'm no expert in human resources management, but this system seems pretty reasonable to me as a taxpayer. Positives include the ability to set individual objectives, the appropriateness of the performance elements, and the presence of a Reviewer who might be able to detect bias in the Rater's judgements. Completing the ACE forms the night before they are due would be a pain, but adult behavior would involve keeping a running record instead.

That said, I personally chaffed at the evaluation systems that I had to endure outside. As a co-founder/co-owner of my software company, I was "above the law" and never subject to formal evaluation. However, I was, and am, subject to pointed comments from any and all of our people. During my academic career at three different universities, the degree of formality in my faculty evaluations has been all over the place. At the end, as Rensselaer became entangled in a corporate model, annual evaluations got progressively more numerical and confining.

How well I could tolerate annual reviews depended on how humanely they were conducted. One department chair turned them into such ordeals that at least one of my colleagues refused to attend. The chair seemed to find everyone's one Big Sin and then beat them about the head and upper body with it until time was up and the next victim came in to face the music. Talk about de-motivating. When another person became acting chair one year, the change was dramatic. My trip to the woodshed that year started with "Tom, you've had a good year." Wow.

In retrospect, I think my personal evaluation system has always been set at some angle away from my formal evaluation systems. As an academic, my core objectives had to be passing along the best old ideas and inventing good new ideas. I would more or less subconsciously track my progress on those two dimensions more or less continuously. Every class hour, every hour spent reviewing the literature on a technical topic, every office hour spent advising a student, and every hour spent noodling in front of a whiteboard would be weighed and often found wanting. Asking me to have a "business plan" for organizing my time made sense to the business side of me, but it was alien to the way I worked as a professor.

Asking me to lay out an annual plan seemed like the wrong thing to do. My real answer would be "I plan to think hard and hope to find an interesting idea on some interesting topic". The fact that the interesting topic might not be anything I'd ever thought about appealed to me but sat cross-wise with custom. The road to academic success usually looks like a long, narrow ditch. You pick a dissertation topic (or, shudder, your advisor gives you one). You turn your dissertation into your first National Science Foundation grant proposal. You submit it a few times until it gets funded. You submit your second NSF grant proposal, proposing to extend the work done in your first grant proposal. Ditto moving from the second to the third. And so forth. After a few years, the in-crowd knows you are a safe bet, and the money trickles in.

In contrast, I took a rather existential approach to my scholarship. I know that what I want to do, what I can make the most progress on, varies randomly from day to day. If the spirit was moving me to wrestle with an equation all day, I would, because I could sense that it was an equation day. If I felt moved to write an exam, or write a paper, or prepare a class, then I would do that, knowing that I'd found my groove for the day and would make the most progress if I just went with it. I can readily recall walking down to the trolley to ride into MIT or Harvard happily thinking "What do I want to think about today?" The problem with this fluid (self-indulgent?) intellectual lifestyle is that it is not career-friendly. It does not mesh well with the engineering school culture of "charge out", i.e., finding "soft money" to pay for some of your salary and all of your graduate students. Engineering faculty are expected to get grants, build up a laboratory (even a virtual one), and support multiple Ph.D. students. Naturally, you cannot get grant money unless you are a known and safe investment, which takes you right back into that long, narrow ditch. Maybe I really belonged in an English department, or was born 40 years too late.

I'd guess that when my colleagues in Math Research were faced with ACE, many had the same thoughts. Indeed, one might expect mathematicians to be even more "in the moment" in planning their mental life than engineers like me. They might believe that informal peer feedback is the most authentic assessment vehicle, especially if they had already reached the GS-15 level and did not want to advance one more step into the Senior Executive Service. For such people, the performance evaluation system at IDA/CCS (see below) might have been much more congenial. Maybe, at heart, they were better suited to being "think-tankers" than "gubmint workers". Or maybe they were just as intellectually self-indulgent as I.

Page 1 of the 18 page ACE personnel evaluation form (see *https://www.icjoint-duty.gov/docs/NSA%20PER%20and%20Guidance.pdf*)

Annual Contribution Evaluation (ACE)
Mid-Cycle Review

Cover Sheet

* CLASSIFICATION: UNCLASSIFIED

This eform is for use by employees who do not have access to ePerformance online via the CONNECT Portal as well as other employees with special circumstances approved by HR.

The Mid-Cycle Review is required of all Employees on a performance plan under the ACE performance management system. It is conducted at the midpoint of an Employee's performance management cycle, and provides a mechanism to assess progress toward objectives and performance elements. The Rater and Employee should discuss the Employee's performance during the first half of the cycle and use this eform to capture narrative summaries of the discussion. No numeric ratings are assigned.

The Rater must complete all of Section 1 and 2 before signing and submitting the form to the Employee. The Employee should then enter comments in Section 3 and return the form to the Rater, who must submit the completed form to HR for processing.

For more information, enter "go DCIPS" in your web browser while on NSAnet.

NOTE: Any field marked with an asterisk (*) is a required field.

Employee Information
- * Name (Last, First MI): Doe,John
- * SID: jdoe
- * Employee ID: 00000999999
- * Email: jdoe@nsa.ic.gov
- Work Role: Analyst
- Pay Level: Pay Band 3

Rater Information
- * Is Rater an NSA employee? ☒ Yes ☐ No (If "No," then Rater's Employee ID and SID are not required.)
- * Name (Last, First MI): Rater,Jane
- * SID: jrater
- * Employee ID: 00000999998
- * Email: jrater@email.gov

Period of Performance
- * Performance Cycle Begin Date: 01-Oct-2008 (dd-mmm-yyyy; e.g. 01-Jan-1900)
- * Performance Cycle End Date: 30-Sep-2009 (dd-mmm-yyyy)
- * Document ID: 2009m00357

PRIVACY ACT STATEMENT: Authority for collecting information requested on this form is contained 10 U.S.C. Sections 1601-1614, 50 U.S.C. Section 402 note, and Executive Orders 12333 and 12968. The Agency's Blanket Routine Uses found at 58 Fed. Reg. 10,531 (1993) as well as the specific uses found in GNSA09 apply to this information. The requested information will be used for promotion, training, assignment and other human resource purposes. Your disclosure of requested information is voluntary. However, failure to furnish the requested information, other than the Employee ID number, may delay inclusion of the information into your personnel files for consideration in human resource actions.

CLASSIFICATION: UNCLASSIFIED

Derived From:
Date:
Declassify On:
Date of Source:

ACE Mid-Cycle Review version 22   Page 1 of 4

Page 1 of the 4 page ACE mid-cycle evaluation form

THOMAS REED WILLEMAIN, PH.D.

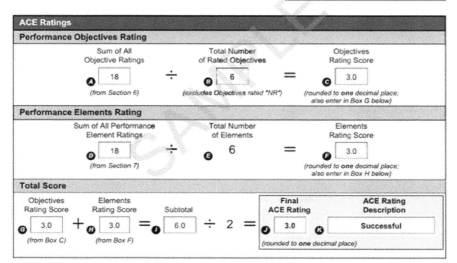

ACE final rating scale and calculations

The Fourth Amendment: A Dangerous Job Killer

Working at NSA and IDA/CSS is always intense, and the technical puzzles we try to solve are daunting, although – guilty pleasure — rather fun because of the high level of challenge. But at least as stressful for me was the requirement that I annually pass an exam to show that I understand how the Fourth Amendment plays out at the operational level.

An unfortunate caricature of NSA is that its people are constantly involved in unconstitutional plots to violate the Fourth Amendment to the Constitution of the United States:

> The right of the people to be secure in their persons, houses, papers, and effects, against unreasonable searches and seizures, shall not be violated, and no Warrants shall issue, but upon probable cause, supported by Oath or affirmation, and particularly describing the place to be searched, and the persons or things to be seized.

My experience at the bottom of the system has been quite the opposite. In fact, the Agency's insistence that I be up-to-date on the legal limitations of NSA surveillance almost cost me my job more than once.

Implementation of the intent of the Fourth Amendment is encoded in Executive Order 12333–United States Intelligence Activities, which, among other things, outlines the charter of the NSA, as well as other directives concerning the Department of Defense. More recently, the FISA Amendments Act has added to the rules governing the Agency.

Early in my college career, I realized that I was not wired for law school. Doing well in high school debate tournaments would not suffice to do well there. Although four of my senior year roommates at Princeton became lawyers, none of their legal inclinations rubbed off on me. As a result, I worried about passing these exams. I was told that I could fail once but had to pass on the second try. If I did not, I could lose my position. As our colleagues at Britain's GCHQ would say, I felt "pressurized". My struggles with these exams also undermined my confidence generally.

The first time through, I had some false hope based on the fact that all those around me in the DARKROOM were whizzing through the computer-administered multiple choice exam, getting scores at or near 100% and casually bragging about completing the exam with time to spare. But when I struggled with the first such exam and failed it, I knew I had a problem. Now, having taken these exams multiple times, I still have trouble meeting the 80% passing criterion and often need the second try to pass.

My excuse is that I think like an engineer, not a lawyer. A central concept is that of a "US person". If you listen carefully to public testimony on this issue, or read official reports to Congress, you will hear that subtly complex phrase, abbreviated "USP". What makes it strange to my ear is that a ship or a corporation can be a US Person. And a ship within 12 miles of the US coast is different from the same ship when 12.1 miles out to sea. And a terrorist located outside the US is a valid target for signals intelligence, but the same person cannot be surveilled by NSA when his aircraft lands at JFK or LAX. When I see the intelligence world described as a "wilderness of mirrors", I think of these annual exams. Luckily, the two-strike rule never brought me down.

There are a number of other exams that have to be passed periodically to be able to see and do certain things. They are all administered from a central web site. At least one time, I had to miss my deadline for re-certification because this web site was down. I hadn't imagined that any NSA web site could fail, but I get the sense (based on this and other evidence) that there is a gap between the research and administrative sides of the Agency. I'm sure that, if we try hard enough, we can find a contractor to blame, abetted by some managers in over their heads.

One summer at IDA/CCS, we worked jointly with ▓▓▓▓ on a project that was unusual in two ways. First, it focused on a topic and on targets that were in the headlines, rather than the usual important but obscure statistical topics that tend to be invisible to the outside. Second, because it was ▓▓▓▓▓▓▓▓▓▓▓▓▓▓▓▓▓▓ ▓▓▓▓▓▓▓▓▓▓▓▓▓▓▓ we had to pass additional tests to certify that we also understood the additional restrictions on data access and use that bound ▓▓▓▓▓▓. (One odd technical restriction was that I could access the ▓▓▓▓▓▓▓▓▓▓▓▓▓▓▓▓▓▓▓▓, could look at the ▓▓▓▓▓▓▓▓▓▓▓▓▓▓▓▓▓▓▓▓▓▓▓▓, but could only ▓▓▓▓▓▓▓▓▓▓▓▓▓▓▓▓▓▓▓▓▓▓ ▓▓▓▓▓▓▓ details. Thus a table listing, say, ▓▓▓▓▓▓▓▓▓▓▓▓▓▓▓▓▓▓▓▓ could not be included in my reports, but a graph plotting the ▓▓▓▓▓▓▓▓ ▓▓▓▓▓▓▓▓▓▓▓▓▓▓▓▓▓▓▓▓ was ok.)

Not surprisingly, and notwithstanding the percentage of my DNA that originated in ▓▓▓▓▓▓, I found understanding the ▓▓▓▓▓▓ restrictions to be even more difficult than understanding our own. (Did you know that ▓▓▓▓▓▓▓▓▓▓▓▓▓▓▓▓▓▓▓▓▓▓▓▓▓▓ including you and me, ▮

▮?)

By coincidence, I actually had to run through the ▮ legalities twice. The first time, I managed to get myself certified to access the data. Then, at "cookie time" (see below) at IDA/CCS, I was introduced to a remarkable Agency mathematician who ▮ Normally, ▮ computer workstation, ▮. Unfortunately, ▮ to use this system to ▮ so I volunteered to ▮ and ▮ This required me be certified (done) and both of us to commit to ▮ (of course, not that they could check – I think!). ▮, he would ▮ we used for this testing ▮

I suppose all this attention to the strict ▮ mission should have been reassuring to me as a citizen. It was, but mostly it was a source of angst ▮

And of course it was a ▮ pretending to be a ▮

This personal tale of woe may not be enough to reassure ▮▮▮▮▮ the Agency, but maybe they will concede that the Agency does ▮▮▮▮▮ all the way down to low level people like me.

Prole to Prince to Pariah:
Blue Badge/White Badge/Green Badge

There is a caste system inside the NSA. It has nothing to do with race, class, or gender. It is all about the color of your badge.

Everyone inside the NSA spaces must wear an identification badge, visible to everybody else. I was warned to never be inside without my badge visible and to never be outside with my badge visible. I was indoctrinated to eyeball everybody I met to confirm that they too had a proper badge visible. If their badge indicated that they required an escort, I had to confirm that they were escorted.

It was never made clear to me exactly what I was to do if I detected a violation of these policies. Presumably I should call in the ninjas. But they might take some minutes to reach my location. Was I then expected to pounce on the unbadged, immobilizing them and perhaps beginning a searching interrogation on the cold floor of the ▮▮▮ building? Fortunately, I never confronted the need for such a confrontation. However, I did gain the ability to quickly scan all around me for a badge check.

I began my sabbatical year as a blue badge. Nothing says belonging like a blue badge. A blue badge is someone who is a full-fledged member of the IC. I believe the CIA folks also have blue badges. So if you watch a TV show or movie that is supposedly set inside the NSA or CIA, don't pay any attention if they don't get this detail right.

A green badge signifies a contractor, or as I came to know them, a "hated contractor" (though nobody else actually said those words out loud). Snowden was a green badge, though the notion of hated contractor predates his treachery. Contractors are resented for two reasons. First, they may be outsiders, even though working inside. (You would think that this definition would ensnare professors on sabbatical, but happily it does not.) Not only are they outsiders, but they are usually paid way more than the blue badgers with whom they work. The second reason is an elaboration of the first. The green badger may actually be a former blue badger who deliberately left the agency in order to come back to work the next week doing the same job for much more money. Hence a bit of resentment.

This strategy of leaving and coming back is not necessarily mercenary. The young woman who was supervising all those resentful sailors forced to "███████████████████████████████████" believed that she had no meaningful prospects of promotion if she stayed in place. But she also believed that if she left and returned as a contractor, she stood a good chance of thereafter re-converting from green to blue badge at a higher pay grade. Though I took her word for it, I never understood the reasoning behind this maneuver. Perhaps when she became a green badge the supervisor class would respect her more because she was paid more. Then seeing how much they were paying her, the supervisors would be happy to have her back at a higher pay grade that nevertheless carried a lower salary than what she made as a green badge.

Someplace along the line, I became a white badge. Foolishly, I gave this no thought. I did miss being "one of the boys", but I was assured that white was a good color to be. With the white badge came the title of "Expert Statistical Consultant" (not quite as lofty as the "Distinguished Visiting Professor" title I got in my previous sabbatical with the FAA, but probably more marketable on the outside). This color change ended up being quite disruptive to my little career in the IC.

It also clouded the interpretation of a little game I sometimes played, just to indulge my sense of mischief. There is a second, weaker caste system inside NSA, overlaid on the first. It is common to almost all tech environments: suits vs non-suits. Most (civilian) working stiffs at NSA are not well dressed and often not too well groomed either. Math Research in particular proudly maintains an average standard of dress a bit lower than that of the rest of the Agency. On occasion, when in the Big Four buildings for SANDBOX I would show up in suit and tie, then prowl the hallways just to watch the reaction. That there was a reaction was the news. Non-suits could care less, but I always found that the suits never failed to check me out. Foolish me, I thought it was that suits wanted to keep track of other suits, who might possibly turn into counter-suits, competing for power and glory. I realize now that it wasn't the suit that made the man: It was the suit combined with that white badge.

My white badge turned out to be a fraudulent white badge, unbeknown to me. One day when I was "outside" at my university office, I got a call from the Agency. This itself was unprecedented, therefore ominous. The inside does not call the outside, at least not at my low level. The call did indeed turn out to be scary, because the caller said that my hiring had been "illegal"! What?! Was I headed straight to federal prison not for improper handling of classified materials, treasonous acts, or botched analyses, but just for being there?

It turned out that the white badge signified that I was a member of some exalted advisory board to DIRNSA himself. That badge therefore made me not just a suit but a Very Special Suit, a Suit with the potential to whisper into an ear and make parts of the org chart vanish. So then a bit of organizational math was activated:

Unknown Suit + White Badge = Serious Target.

How did this come to be? A bureaucratic foul up by a woman who was apparently too powerful to suffer for her mistake. Let's call her HELEN. HELEN had taken a number (I think about eight) of academics who worked part time for NSA and placed us in a certain ███████████ associated with white badges. This was a ███

There were some warning signs that things were not right before that fateful phone call. During my sabbatical, there seemed to be a problem with my paycheck every month. We later discovered that HELEN ████████████████████████████████
When I had my first meeting with LUKE, the head of the Research Directorate, he listened to the problem and stunned me (but not my Math Research colleagues) by shaking his head and wishing us luck with the problem. Huh? This guy was the civilian equivalent of a Major General and he wouldn't even try to fix this? HELEN must be a Force-To-Be-Reckoned-With.

When some eagle-eye eventually discovered the White Badge Eight in the wrong budget line, the Agency had some options. They could deal with HELEN (my favorite) or just quietly turn us into green badges. Instead, the Agency fired us all.

The problem with this option was that some of the others in the White Badge Eight were pretty heavyweight guys, well-connected and well-respected within Math Research. This mess would not help Math Research's liaison with the outside, nor would it help Math Research deal with its overfull analytical agenda. So, to his eternal credit, MATTHEW, then the chief of Math Research, found a way to keep us around. His solution, in my case at least, was to turn me into a hated green badge and broker a placement elsewhere in the IC. In my case, I was fortunate to be selected by IDA/CCS in Bowie, MD for work on SCAMP projects (see below). I could still use my green badge to get into Ft. Meade (and eventually the place of exile of Math Research in Laurel, MD), but I and the others were basically booted off post.

I'm willing to assert that, by now, our new affiliations have gotten their money's worth. In my case, the requirement for each member of the SCAMP teams at IDA/CCS is that we produce one report before returning to the outside; I have always produced at least two.

Anyway, I had completed my odyssey from blue badge to white badge to green badge. The NSA green badge carries no weight and literally opens no doors at IDA/CCS, so in that sense I had become a no-badge.

My new badge at CCS, ironically, is white. It carries a prestigious-sounding low (two-digit) number, but the prestige wore off a bit when I discovered that there is somebody else at the same place who also wears badge ■. Unlike NSA badges, IDA/CCS badges cannot be taken off the premises. The guards got confused when I showed up for work and request badge ■, so I just told them to give me the ■ with the handsome picture on it. They would chuckle... and give me the other one.

Booted to Bowie:
The Big SCIF

It turned out that IDA/CCS in Bowie, MD is a candy land for geeks. Wikipedia provides the dry, historical description:

> The Institute for Defense Analyses (IDA) is an American non-profit corporation that administers three federally funded research and development centers (FFRDCs)—the Systems and Analyses Center (SAC), the Science and Technology Policy Institute (STPI), and the Center for Communications and Computing (C&C)—to assist the United States government in addressing national security issues, particularly those requiring scientific and technical expertise. It is headquartered in Alexandria, Virginia...
>
> IDA's support of the National Security Agency began at its request in 1959, when it established the Center for Communications Research in Princeton, New Jersey. Additional requests from NSA in 1984

and 1989 led respectively to what is now called the Center for Computing Sciences in Bowie, Maryland and to a second Center for Communications Research in La Jolla, California. These groups, which conduct research in cryptology and information operations, comprise IDA's Communications and Computing FFRDC.

When I asked about CCS before landing there the first time, someone told me that CCS solved the problems that were too hard for the NSA. I thought that was a somewhat prideful statement, but took comfort in thinking that I had not been demoted to the junior varsity. Now that I have spent several summers there as a member of the Adjunct Research Staff, I have a more balanced view of the relationship between NSA and CCS. CCS's budget ███████ There is a ███████ staff at CCS that ███████ agenda. ███████ injects a few ███████ for periods of ███████ so NSA does not really completely hand over its most difficult problems to CCS.

Here's an aerial view of CCS. The entire building is a SCIF (Secure Compartmented Information Facility). A SCIF is a space to speak of secret things without fear of being overheard. If the V-shaped design reminds you of a stealth bomber, so much the better, but I'm sure it's just a happy coincidence. Note the sundial (discussed below) inside the U-shaped grassy area between the building and the parking lot. The picnic tables (also mentioned below) for lonely lunches are in shadow by the fence line behind the building. Not visible are security measures, an underground level gym, and possible storage facilities for alien computing technology. As far as I can tell, the trees are actually trees, but one never knows: Everybody knows by now that some innocent-looking rocks are not just innocent-looking rocks.

IDA/CCS: One comfortable SCIF (credit: Google Earth)

SCAMP: Mixing the Monastic with the Secular

The short term activities at CCS are tightly coupled ▮▮▮▮▮▮▮-▮▮▮▮▮▮▮. The short term projects are called "SCAMPs". There are usually two or three SCAMPS every summer, running for 12 or 13 weeks. It is tempting to believe that "SCAMP" is a contraction of "summer camp". It is not; there is an official name that contracts to SCAMP (Special Cryptologic Advisory Math Panel), but, hey, it really is a summer camp — for propeller-heads, code-monkeys, and other uber-geeks. (IDA prefers to refer to the SCAMP groups as "tiger teams" on its web site.)

The first SCAMP was held in 1952. The report of Stewart S. Cairns on the 1953 SCAMP shows why SCAMP was created:

> The purposes of the project known as SCAMP, which held its first session in the summer of 1952 and its second in the summer of 1953 are: a. To support research in computational aspects of discrete problems related to the interests of the National Security

Agency. b. To educate a group of competent cleared mathematicians in mathematical problems and techniques applicable or potentially applicable to NSA problems.

Mathematical interpretations of certain practical problems of the NSA are closely inter-related with a number of currently studied research questions in pure mathematics. The relationships are so close that, on the one hand, mathematicians unacquainted with NSA problems may well develop methods contributing to their solutions and, on the other hand, mathematicians working directly for the Agency may achieve results of pure mathematical interest. It was in the hope of capitalizing on this situation that the NSA undertook to sponsor SCAMP. The Agency stands to benefit not only from the immediate products of SCAMP, but also from the creation of a pool of cleared, informed and highly competent mathematicians who may later render valuable service.

Another merit of the SCAMP program is the opportunity for NSA mathematicians to associate professionally with outsiders working on related problems. In a recent discussion, Dr. J. Weyl of the ONR [Office of Naval Research] drew a contrast between "monastic and secular" mathematics, which fits the present situation much better than the usual dichotomy into "applied and pure" mathematics. To some extent, Agency personnel must lead a monastic life. For the sake of their effectiveness, however, as well as for their personal and professional welfare, it is extremely important to provide for the type of outside collaboration involved in SCAMP, thus lessening the dangers (1) that they will pursue their own Agency research without capitalizing on closely related "secular" mathematics and (2) that their work will be more or less stultified for lack of the stimulation which comes from a wide variety of professional contacts. As already mentioned, the important mathematical problems of the Agency are closely intertwined with unclassified current research.

["Report on SCAMP 1953 to the Director of the National Security Agency" by Stewart B. Cairns. *https://www.nsa.gov/news-features/declassified-documents/friedman-documents/assets/files/panel-committee-board/FOLDER_353/41717879075719.pdf*]

(Note that much the same motivations can be ascribed to the Math Research sabbatical program that got me from outside to inside.)

NSA people are represented on SCAMP teams, ██. The rest of the teams are a mix from three groups: Always some members of the outstanding permanent CCS research staff, several academics like myself from the adjunct research staff, and a number of carefully selected and always awesome graduate students.

The graduate students do not usually return for a repeat SCAMP; instead, they go back to complete their dissertations. After graduation, a very few are invited to return to CCS as permanent staff. Others end up elsewhere in the national security establishment, at Lincoln Labs, Sandia or similar organizations. The rest just go back outside. In this, they resemble Math Research sabbatical visitors, who mostly go back outside with the implicit ongoing mission of being quiet ambassadors for the Agency.

My impression of the graduate students I worked with is very positive. I was surprised to see that, like me, they often come with no prior knowledge of the problem area of their SCAMP. What they have is big brains and strong general backgrounds in math and/or computing. As one of the few statistical people in most SCAMPS, I have often tutored these kids in both statistical theory and statistical data analysis using the *R* computer software system. It never took them long to get up to speed. Unlike me, they usually created social lives for themselves, but sometimes they were able to outwork me too. I remember one young man from the University of Illinois who was always there in the office next door when I came in on a Saturday or Sunday and still there when I left. We were working independently on the same problem. I would say that I did some useful work and got results that complemented his, but let's face facts: If we had both been in contention for a prize, he would have won it.

Loosey-Goosey

Life in CCS can be very good compared to life at NSA. The entire building is a SCIF, so conversation is much freer than at NSA. You can see SCIFs depicted in movies, in which the heroes stuff themselves into a cramped, airless vault within a vault to say things they cannot say elsewhere. Bowie's SCIF is big and airy, full of sculpture and art, chess boards, white boards, and soft furniture. It is the King of SCIF's.

More than that, and even more so than NSA Mathematics Research, Bowie is more academic than the academics. The management culture is to stay out of the way, so annual performance reviews were reported to me to consist of short, vague conversations consisting of statements like "I think I'll keep working on X. I got somewhere on the problem this year. I might have good results in a few more years." The management reaction might consist of a request to hear about the early results and maybe make a few esoteric suggestions about future lines of investigation. If that isn't research Nirvana, nothing is.

I cannot imagine a university professor having such a conversation with a department head or dean. Especially in engineering schools now, the conversation would focus on grant money, committee assignments, grant money, teaching assignments, and grant money. The professor would be given a weighted numerical performance score based on his or her grant money, number of graduate students launched out the door, course evaluations completed only by students nurturing a grudge, and... grant money. The interview would conclude with a demand for a plan for getting more grant money. (Maybe it was good for me to retire a few years ago, before the bitterness and disillusionment set in too deep.) I have no doubt that the contrast between the lives of new Ph.D.'s at Bowie and at universities works in favor of Bowie's recruiting process. It might even balance out any vague worry about selling one's soul by accepting "DoD blood money" and supporting the "evil NSA".

Cookie Time

To complete the picture of the work environment at CCS, consider "cookie time". Every day at 3 pm, all resident nerds are expected to leave their burrows and join in the cookie time event in the large common area. A cart is wheeled in full of cookies, potato chips, and sometimes Cheetos. The cart is mobbed. Conversation erupts around deep and obscure technical topics: ▮▮▮▮▮▮▮▮▮▮▮▮▮▮ etc. The food lasts maybe five minutes. Some of the conversations last maybe 2 hours. Over in the corner, two-person teams are playing Bughouse, a scary variant of chess in which pieces taken from one board can be injected into a second board. (I don't recall many times when two people just played a conventional game of chess. I was terrified that someone would ask me to play Bughouse, so I always kept my distance. I don't think there is enough Aricept to ramp me up to a respectable level of play at even old-fashioned chess. And without even the ability to juggle, which seems to be very common among Ph.D. mathematicians (including Claude Shannon), I expect I would have been exposed to all as a fraud unworthy of space in Bowie.)

The administrative staff made it clear that they considered cookie time a serious burden. The common space always was a mess afterwards. (When one exists in the higher realms of thought, one should not be distracted by the state of the carpet, should one?) When I protested that the fare was uncongenial to diabetics and should include some sugar-free cookies, I got a blast of complaint in answer. I didn't even mention that there were no drinks provided. I would have to go downstairs to the first floor to get a cold soda from the vending machine; on the other hand, it was the cheapest soda on the East Coast: $1.25 for a diet Coke and, sometimes, the best choice of all, diet Dr. Pepper.

Despite the irritation of those who make it happen, cookie time succeeds in promoting both human and intellectual contact. I'm sure, especially with the fiscal discipline shown in limiting the volume and variety of snacks, that the ROI on cookie time was the highest anywhere inside DoD.

I confess that I often did not realize when cookie time came round and would therefore miss it entirely. I also tend to be a bit misanthropic when I'm wrestling with a problem and losing, so I often skipped cookie time on purpose, to spare my colleagues the grumpy old man act.

I have only been at CCS once outside of a SCAMP. That week during the fall made me think of CCS like the little tourist town of Port Clyde, Maine. Most of the year, most of the office space at CCS is empty. The gym in the basement is never overcrowded, cars can be parked close to the entrance, and it is possible to grab more than one handful of chips at cookie time before the supply is gone. But during summers, the summer people overwhelm the place, filling the parking lot with rental cars, stuffing every empty office, taking all the lockers in the gym, crowding the Bughouse table, plundering the cookies, and bothering the staff responsible for keeping all the computer systems up.

The Real People of IDA/CCS

The people one meets in Bowie are no less interesting than those one meets at Ft. Meade. They do seem a little different as a group, as befits their different, often longer-term, focus.

<u>SEAN</u>. SEAN runs the place with the kind of light touch that makes CCS so attractive. I've had only a few personal meetings with SEAN; I got the impression that his mind was elsewhere when we met. Whether this was down to the way he "inherited" me or his weighty responsibilities was not clear. When I asked him to just let me work on things that I considered important, he was not able to give me that latitude. Disappointing, but perhaps understandable in that he has to respond to his sponsor's agenda, not take a chance on mine. Every week, each SCAMP team gives a public update in the only meeting that is considered "required"; whenever possible, SEAN shows up and asks pertinent questions. I like that kind of behavior in a leader. Others in similar positions, both in the IC and in academia, let themselves get stuck in the budgets and other management issues and lose touch with the "real work".

THEODORE. THEODORE has since retired, but his memory lives on as the Indispensable Man. Two or three people now do what THEODORE did to organize all the SCAMP logistics as well as make his own technical contributions. I learned early on that THEODORE was the key to customizing my work environment. When I had a too-noisy set of roommates (typically three or four of us to a room), he found me wonderful ear protectors under which I could zone out (and miss cookie time). When I was allergic to a room, he found me another to use. I suspect his get-it-done gene grew from his time as a submariner, where every detail has to be just right, or else. The national security establishment is blessed to have people like THEODORE who keep everything running.

VERNON. Sadly, I returned one summer to find that VERNON had passed away at too young an age. The summer before, knowing his prognosis, he was nevertheless working in his office, and I had made a few visits just to offer support. In those days, VERNON's parents would visit. His dad was holding it together, but his mom was melting with the sorrow of it, and VERNON told me he could handle everything but that.

Against stereotype, may of the CCS staff have high verbal facility, and VERNON was surely among them, so he was a pleasure to talk with at cookie time. When we met, I was in a SCAMP devoted to ███████████████████ and VERNON would assail me with discouraging arguments about why ███████ ████████████████████████ I would respond that it still made sense to ████

My most treasured mental image of VERNON is a cookie time image. VERNON was slouching against a wall, with one hand curled around a handful of potato chips. We talked and talked. Half an hour later, I noticed something: VERNON was so intent on the conversation that his hand was still where it was at the beginning and, remarkably, still curled around the phantom potato chips that were no longer there. How's that for concentration beyond the demands of the body! To regular people, that would make VERNON a strange guy. To me, it made him a god of the intellect.

ZACHARY. ZACHARY is a ▮▮▮▮▮▮▮▮ math Ph.D. with a long record of secret accomplishment. While I've never had occasion to engage him in a technical discussion of my own topics, that probably means I've missed a good bet. But I have enjoyed his company at many a cookie time; he's one of those open people who don't mind a new person wandering in to his conversational circle.

As with VERNON, I carry an instructive visual memory of ZACHARY. ZACHARY dresses appropriately for his position: baggy pants and one of only two colors of polo shirt haphazardly thrown on (i.e., he is my sartorial peer). One day, several of us engaged him in an interesting chat about ▮▮▮▮▮▮▮▮. ZACHARY held forth at length, which is to say, for many handfuls of Cheetos. By the time he was finished, enlightenment reigned all around, and several of us had our shirt fronts dusted with orange Cheeto residue. We walked away unaware and unconcerned about the dusting. Try that at Morgan Stanley. We taxpayers are lucky to be paying the salaries of people like ZACHARY, for whom results matter and form counts for little.

ALPHONSE. (ALPHONSE has an interesting real name. Maybe I've not met enough people in my life, but I must say that I've encountered some utterly unique (real) names in the IC. I sometimes wonder if little geniuses get recognized early and are then given names unlike all us regular kids.) ALPHONSE is one of the ▮▮▮▮▮▮▮▮ at CCS and anchors an entire area of mathematics. He is also a serious artist, with his ▮▮▮▮▮▮▮▮▮▮▮▮▮▮▮▮▮▮▮▮▮▮▮▮▮▮▮▮▮▮▮▮▮▮ decorating not just the Bowie SCIF but also ▮▮▮▮▮▮▮▮ His work ▮▮

Always floating a little distance from reality, I had not understood the meaning of ALPHONSE's sculpture. It had to be pointed out to me that the big thing was, in fact, a sundial. Until that moment, my interpretation had been a projection of my struggles to solve my SCAMP problems. For many summers, I saw the sculpture as a cloven rock. I thought the huge rock represented the tough technical problems I was beating my gums against. I thought the huge metal right triangle embedded in the middle of the rock represented the Power of Mathematics to Vanquish the Problems. Not a bad interpretation, but not correct.

But then, I think it wasn't until I was in my 60's that I realized that the expression "A watched pot never boils" did not mean "Keep watching and you can keep the pot from boiling, because boiling water is dangerous". My family constantly wonders at my delayed insights. They've given up being worried by them and just find amusement. Secretly, they may wonder how I can get anything useful done in the Real World.

<u>GARTH.</u> GARTH is someone I never worked with, but I saw him every summer during the required weekly team progress reports. GARTH is an expert cyber warrior and a veteran of military intelligence, which is probably where he learned his special craft. Most of the time, I have little or no idea about what he's talking about. What I can appreciate is the way he talks about it. GARTH has got to be the best briefer on the planet.

Most of the people who get up to present their team updates lack the essential discipline required to do the job well. They commit the usual sins associated with having a Ph.D., especially (ironically) if they are also professors. That is, they do not take the time to build an organized flow into their remarks, do not work to bring the technical level down to match their audience, and never, ever make their remarks fit into their allotted time. On the other hand, GARTH's briefs are crisp, clear and timed perfectly.

I think I've spoken only one time to GARTH. I once stuck my head into his office door to tell him he was the best briefer I'd ever heard. He looked startled and mumbled thanks. Five seconds on, when I'd disappeared like a ghost, he must have wondered who the hell that summer person was.

BYRON. BYRON is one of the "old pro's" on the permanent research staff. Most summers I'd had no dealings with BYRON, just seen him around in ever-present shorts and tee shirt. I knew he and his wife always served as a host family to lonely college students away from home for the first time (a great thing, as I learned from my nephew at the ███████████████████████), but beyond that BYRON was just another pretty face.

Then, in what turned out to be my final SCAMP, I stumbled into a project that may have been the most impactful I've done at Bowie. BYRON was the leader and showed a scary-deep knowledge of computing at a very ███████████████ level. (Not that I should have been surprised at that, since that's a common denominator among the permanent research staff: Their expertise ranges across all aspects of computing hardware and software.) In a perfect mix of old and new, EDWIN, a brilliant summer grad student, teamed with BYRON to achieve a technical feat that was considered to be impossible. SCOTT (see below) joined me and new-guy EVAN to nail down the statistical side of the work, while people from NSA's elite ████ group kept tabs and provided support and encouragement.

The final report of almost every other team in our group documented their failure to crack their problem. They had attempted an extremely difficult task, taken many runs at it from different angles, never quit, but never found success. I had never seen a SCAMP problem resist so well the efforts of so many capable people. However, their effort was not wasted. A lot was learned about that problem, and a list was compiled of things that seemed promising but turned out not to work. The problem won't go away and will have to be attacked again, but the next kamikazes will know what not to try.

In contrast, our group of BYRON, EDWIN, SCOTT, EVAN and me were excited to report a stunning success (actually, two and a half successes). As with the work I did at NSA, it is usually difficult to know what impact the research at CCS will have. Not in this case. Our work generated immediate interest. ███████████████████████████████ I'd never seen *that* before. Quite a feeling, even knowing that my part of the project was secondary. What a triumph for BYRON and EDWIN. BYRON remains at CCS, a positive example of Your Tax Money at Work. EDWIN went on to work in another part of the national security apparatus. Another positive example of Your Tax Money at Work.

The itinerant professors. One of the groups that show up as summer people in Bowie are the professors on the rolls of Adjunct Research Staff. The cast of characters changes every summer, since CCS's sisters in Princeton and La Jolla compete for many of the same people and offer "certain lifestyle advantages".

A tiny subset of the summer professors brings in much-needed statistical expertise. When ISAAC was around, it was always instructive to look over his shoulder because he would be exercising the latest in exploratory data analysis, including wonderful data graphics.

When SCOTT was around, we would often scramble together to get some data to work on. When we had data, he was always a great partner, fixing any errors I made, letting me fix any of his, and complementing my simple analyses with more rigorous alternatives. SCOTT is also an expert on the early days of the business, starring Alan Turing at Bletchley Park, and a source of many an interesting tale.

Sometimes, no other statistically-inclined faculty show up, and I feel the lack of somebody with whom I can just kick around half-ideas. If GRAHAM is in the SCIF I have what I need, and if LINCOLN is around I have more than what I need. Otherwise, poor JASPER has to provide attention and whiteboard space. JASPER (who is reputed to occasionally *not* wear shorts, tee shirts and sandals, though you can't prove it by me) is a CCS staffer who is patient enough to let me interrupt his own work, which focuses on advanced forms of machine learning, and articulate my ideas on such things as stochastic processes or Monte Carlo simulation, which are not his specialties.

That I have to bother JASPER at all points to the need for even more statistical expertise at CCS. Most of the projects I have been involved with were, at their core, totally opaque to me. As a non-Unix/non-code monkey who works in Windows and programs only in R, I am a definite outlier. Now and then I suspect from his reaction to my questions that he wonders who let me through the door. But even though I always know next to nothing about the computer science background of a problem, I have always found a way to contribute because there is always some statistical aspect that needs to be addressed. It would be useful to insure that future SCAMPs always include additional expertise in statistics and data analysis.

A few of the Adjunct Research Staff are not just summer people. Some who 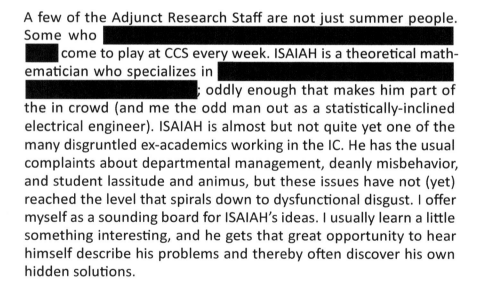 come to play at CCS every week. ISAIAH is a theoretical mathematician who specializes in ▓▓▓▓▓▓▓▓▓▓▓▓▓▓; oddly enough that makes him part of the in crowd (and me the odd man out as a statistically-inclined electrical engineer). ISAIAH is almost but not quite yet one of the many disgruntled ex-academics working in the IC. He has the usual complaints about departmental management, deanly misbehavior, and student lassitude and animus, but these issues have not (yet) reached the level that spirals down to dysfunctional disgust. I offer myself as a sounding board for ISAIAH's ideas. I usually learn a little something interesting, and he gets that great opportunity to hear himself describe his problems and thereby often discover his own hidden solutions.

This listening role is quite conventional for me. Over the years, I have helped many a colleague or doctoral student solve their own problems simply by letting them listen to themselves describing their problems. Though I am allergic (physically and psychologically) to all animals, I call this my "woof woof" role: I just have to sit quietly and say "woof woof" now and then like a good dog while the other person solves his own problem. In turn, besides learning a few bits of exotica, I get ISAIAH's feedback on my ideas and take some inspiration from his deep attachment to the game of solving hard problems.

There is something else I get from ISAIAH and GRAHAM and HENRY and LINCOLN and COLIN and all the others: A sense of what is at stake. The work we do might look to some people like self-absorbed play for overgrown kids and navel gazers, but it is actually a weapon. I have frankly been stunned to never once having heard folks in the IC talking about the multitude of threats to the US and our role in the invisible fight. There are no pep talks, no locker room rah-rah. This is especially true at CCS where, unlike at Ft. Meade, it is rare to see anybody wearing cammo. But I can sense without a doubt that the impetus to try harder, think deeper, and try just one more angle is right below the surface. Maybe keeping it all inside works to make it stronger, but it's there.

What we have here in the US, what we are trying to build and grow, our notions of freedom and decency, are worth protecting. And nobody has the right to kill my grandchildren just because they're Americans. Nobody. So it's not just math, it's math with a mission. Math as a weapon.

The Women of Bowie

(Women are also "real people of CCS", but they get their own chapter.)

The IDA web site breaks down the staff at the three sister centers by discipline, but I have not seen a breakdown by gender.

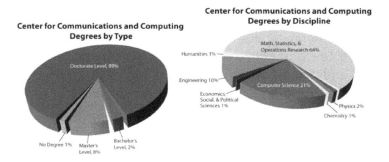

Technical backgrounds of IDA FFRDC professional staff (*https://www.ida.org/en/IDAFFRDCs/CenterforCommunications.aspx*)

Allowing for the possibility that I may be missing something, I would argue that those pie charts get most of the story because academic discipline counts for more than gender at CCS.

Though women are still a clear minority at Bowie, they are not invisible. Women make up the majority of the administrative staff. They do a good job. The admin side of Bowie is noticeably tighter than that at Ft. Meade or even at the mother ship of IDA headquarters in Alexandria, VA. The number of women among the permanent research staff is increasing, most noticeably among the bright graduate students who show up for SCAMPS.

Several of the new female research staff hires have added, among other things, a new level of sociability to a group that tends toward classical geeky introversion. That made for a different lunch dynamic. One of my problems with Bowie is that there are no eating facilities. There are two picnic tables in the back yard for those packing their own lunches, but the Maryland heat and humidity and an irritating drone from the HVAC systems often made me, some ants and a few spiders the only entities out there. Most folks take off to eat "on the economy", but this usually wasted an hour, half of it burned just driving back and forth to a mall. Those staying in to eat would gather in the common space at a few round tables that could accommodate up to a dozen people. With the coming of some new women mathematicians, those tables became more sociable, and I tended to join in when I wasn't feeling too bummed about progress on my SCAMP problem. It might seem impolitic to say that women "humanized" the place, but I've always thought it was more interesting to talk to women than men. Men I understand; women I may or may not understand, but I can recognize top talent, and this group of women has it.

<u>HOPE</u>. HOPE was the leader of one of my SCAMPs and did a great job of keeping everything on track. In between, she kept up her own technical contributions while massaging some egos, teaching me a few things, and getting ready to ███████████████ She also organized and invited me into a "reading group". Reading groups represent the process of continual advancement that one finds in any elite research group, whether on the outside in a university or inside the IC. I was pleased to take my turn introducing the group to work I was doing at Smart Software on time series analysis.

ANNABELL. ANNABELL works her projects with an intensity that I admire. She usually walks around with a scowl too. If she were a man, her intensity and distracted scowl would win her points, so we'll let her keep the points as well-earned.

I've had only two notable interactions with ANNABELL, both involving some friction that turned out to be unnecessary. The first one was literally on the first day of my first SCAMP. ANNABELL graciously invited me and one other newbie to a discussion of an interesting technical problem that I'd never heard of called a "███████████████", which attempt to ███████████ by ███████████ (ANNABELL and the others in the room were polite enough to suppress their surprise that I'd never heard of a ███████████████████████. What I thought of as a big personal challenge might have looked to others like one honking big risk.) I felt drawn to this problem since it tapped into the long-suppressed electrical engineer in me. However, I had already announced my intention to work on another problem, and I wrongly assumed it would be disloyal for me to bail on day one. At that point I started to think of ANNABELL as some sort of evil technical seductress, luring me away from my duty. Later, I learned that the culture of SCAMP is much more free-wheeling. I'd invented an imaginary loyalty issue: the cultural norm was that anybody can work on any of the simultaneous SCAMP problems, and restless people end up working on more than one simultaneously. From feeling suspicious of ANNABELL, I'd gone to avoiding her out of my own feeling of guilt about possibly disappointing her. Meanwhile, all of this psychological turbulence was doubtless invisible to ANNABELL, who was too busy protecting the country to notice.

My second interaction with ANNABELL was more public. Some summers later, we found ourselves on a team working a new SCAMP problem having to do with ███████████████████████. About this time I became aware of a ███████████ command called ███████████

███████████████████████████████████████. I realized that this command would be a wonderful source of real-time data to use in my introductory statistics class back at Rensselaer. So at night I would play with ██████████ commands and build in-class demonstrations and out-of-class group projects around that data. In our weekly team meeting, I mentioned that I was doing this and found it interesting. ANNABELL unloaded on me in front of the whole team, accusing me of jeopardizing the security of our project. I took her point, since I had been told that sometimes it was not the ██ On the other hand, I had also been told that both ███████████████████████████ I assumed then that they must ███████████████████████████████, so — perhaps rashly — I made the further assumption that I had good cover for these ████████████████ experiments. In any event, this was the only time I was really steamed when working on the inside. The episode passed with no further discussion, and ANNABELL and I had no further contact during that project.

Since there were so many working on that team, the lack of interaction was not unusual. In subsequent summers I've made a point to say hello to ANNABELL when I see her in the halls. Maybe there's time for one more chance to share a problem with ANNABELL; there is a lot she could teach me. Still, I sense that she's a cat person, and that might be a problem for me. (OK, yes, I have issues with cats, but, hey, I do like all human babies, every single one.)

<u>KATHERINE</u>. KATHERINE is the Queen of Computing and one of those rare people whose patience matches her technical skills. She supervises a small staff of hardware and software experts who share her patience for people like me who tend to have noob-level questions.

THOMAS REED WILLEMAIN, PH.D.

Here are two recent job postings that give a good sense of the skill level of IDA computing staff. The first happens to be at the Princeton facility, but could have appeared at any of the three centers. Notice the wide range of skills required. Not many of those people hanging around on street corners. My guess is that the toughest requirement would be that one about coaching folks who already consider themselves experts.

Research Programmer /Mathematical Applications

Location: **Princeton, NJ**
Department: **Center for Communications Research Princeton**
Closing Date: **3/31/16**

Description

Overview:

Under general direction, this position performs scientific programming projects that require mathematical maturity, understanding of modern computer architectures, experience with mathematical software packages, and classified domain knowledge. Individual will be the in-house specialist in code optimization, help research staff improve their programming skills, and improve, optimize, and maintain a complex library of subroutines. Incumbent will also support our high-performance computing system administration effort.

Responsibilities:

- Optimizes the mathematical algorithms in the CCR software library across varied computing architectures, and assists research staff to contribute new software to the library.
- Achieves and maintains expertise in all computer languages, math libraries, and mathematical software packages in use at CCR.
- Coaches individual research staff members to improve their programming skills.
- Monitors the high-performance computing environment as seen by research staff, and makes or recommends changes to our programming tool set.
- Documents our programming tool set using CCR's internal wiki, technical reports, code examples, and videotaped technical talks.
- Assists research staff members with particular code debugging and optimization problems.
- Becomes familiar with the technical aspects of communications protocols and signal processing as needed to support research projects.
- Maintains expertise in numerical libraries and solvers.
- Is intimately familiar with parallel programming tools and paradigms in use at CCR.
- Stays current with outside progress in software and software development tools.
- As directed, writes special-purpose programs and utilities.
- When required, assists with system administration of CCR's high-performance computing network.
- Performs other duties as assigned.

Qualifications:

- U.S. Citizenship is required.
- B.S. Degree in Mathematics.
- One year or more of directly-related work experience.
- Knowledge of mathematical algorithms, analysis of algorithms, code optimization, and computer languages, equivalent to a B.S. in Computer Science.
- Competence using programming languages such as C, C++, and Fortran 90 for mathematical applications, and higher-level languages like Perl, Python, Ruby, and Lisp for prototyping.
- In-depth knowledge of parallel computing, high throughput computing, MPI, OpenMP, Posix threads, and either CUDA or OpenCL GPU programming.
- Familiarity with commercial and open source math packages such as Matlab, Octave, Sage, Maple, GMP, Magma, Mathematica and Macsyma.
- Excellent interpersonal skills and ability to communicate effectively both orally and in writing with all levels of personnel.
- Ability to teach effectively.
- Ability to work evenings and weekends to meet the needs of the organization.
- Ability to obtain and maintain necessary security clearances.
- Ability to attain within one year and maintain security certifications required by DoD Dir 8570.

IDA is an equal opportunity employer committed to providing a working environment that is free from discrimination on the basis of race, color, religion, sex (including pregnancy and gender identity), sexual orientation, national origin, age, disability, status as a protected veteran, marital status, genetic characteristic or any other legally protected condition or characteristic.

IDA/CCR job posting for research programmer

WORKING ON THE DARK SIDE OF THE MOON

The second is a job posting for a systems programmer at Bowie. The bit about responding to user's requests "in a professional and timely manner" is gold to me, though it might not stand out to an applicant.

Sr. Systems Programmer - Bowie, MD

Location: **Bowie, MD**
Department: **Center for Computing Science**
Closing Date: **2/29/16**

Description

Overview

Under general supervision, this position provides for the various elements of continual operations of the CCS computer systems. The incumbent will install and maintain various operation systems on servers and work stations, working to analyze, update or suspend systems as necessary. Incumbent must have strong analytical ability in order to diagnosis and correct system problems. This position supports IDA's Center for Computing Sciences Information Technology division. No supervisory responsibility. No budgetary responsibilities.

Responsibilities

1. System Support

- Manages and maintains computer systems (e.g. servers and workstations) on multiple networks by monitoring system performance, configuration, maintenance and repair.
- Develops new system and application implementation plans, custom scripts and testing procedures to ensure operational reliability. Trains technical staff in how to use new software and hardware developed and/or acquired.
- Performs software and hardware installations.
- Provides in-depth internal UNIX kernel support.
- Performs troubleshooting as required. As such, leads problem-solving efforts often involving outside vendors and other support personnel and/or organizations.
- Provides performance monitoring, tuning, and capacity planning.
- Depending on the system, may perform systems administration functions, such as adding user accounts, performing system accounting and other related activities. Support includes error detection, analysis and timely intervention.
- Provides innovative, practical solutions to a wide variety of hardware and software problems.
- Stays current with technological developments in systems administration technology and recommends ways for PCC to take advantage of new technology.

2. User Support

- Interfaces with vendor's customer support personnel in resolving user problems.
- Responds to users' requests, providing professional support in a timely manner.
- Maintains a high level of expertise in the area of focus (e.g. Linux and/or Windows).

3. Perform other duties as assigned

Qualifications

- BS/BA degree in Computer Science or related area, with six or more years of progressively responsible work experience with computer systems including Windows and/or Linux operating systems.
- Demonstrated experience in scripting and programming.
- Demonstrated experience with various hardware and software.
- Knowledge of networking
- Excellent written and verbal communication and interpersonal skills.
- Ability to obtain and maintain TS/SCI Clearance with Fullscope Polygraph; Active TS/SCI Clearance with Fullscope or CI Polygraph, preferred.

IDA is an equal opportunity employer committed to providing a working environment that is free from discrimination on the basis of race, color, religion, sex (including pregnancy and gender identity), sexual orientation, national origin, age, disability, status as a protected veteran, marital status, genetic characteristic or any other legally protected condition or characteristic.

IDA/CCS job posting for senior systems programmer

The computing facilities at the Center for Computing Sciences are as good as one would expect them to be at a place with that name. I think my favorite is the extensive backup facility. More than once, I've had to run to KATHERINE or her people and ask them to restore some critical source code that I had managed to accidentally delete. The dream of getting it all right the first time grows more elusive with age, experience notwithstanding.

High Level Security: What's a Happy Hour?

CCS, unlike NSA, does not have its own ninjas. It has technical protections and a guard force that seems on a par with what I see at the GE Global Research Center down the road from my home. What it does have, among other assets, is a security office run by two women who do not look like super-cops but do the job in their own competent and idiosyncratic way.

One of my most pleasant memories of summers in Bowie is the sound of rollicking laughter pouring out of the security office. What was so funny? I've forgotten, but surely nothing related to work. Given all the nasty threats that these women have to deal with, their sense of humor was a rare but healthy way to deal with the tension. Laughter being infectious, I always tried to find some excuse to pop my head in and get a good dose before trundling off to my workstation. Laughter: A definite workplace plus, right up there with the cheap Diet Dr. Pepper. For some reason, not mentioned on those job announcements.

One of the security team's main challenges is dealing with the intemperate behavior that emerges when all the researchers are forced into the one auditorium every summer and made to endure the annual computer security briefing. This is somewhat analogous to confession in my Catholic church: You have to do it at least once per year; everyone knows that he or she is weak and would benefit from doing it; you can't just let it be done to you but have to co-produce it; and it can be mass-produced using general absolution. These annual briefings are our general absolution.

The big stick is the threat of losing the security clearances that are the breath of life in the IC. But Bowie offers twin inducements to make it all easier to swallow, literally. One is the chance to assemble our very own sandwiches from a long line of raw materials (ham, cheese, turkey, mayo, mustard, etc.) and to choose among a set of cans of soda. Posh, right? So relax, Mr. and Mrs. Taxpayer: it's a cheap lunch. But don't look too closely at what happens when everybody gets inside the auditorium and the group builds up a critical mass for what GCHQ might call "laddish behavior".

One infamous summer, the main briefing on security was presented by two women who came down from Ft. Meade to address the barbarians. They had a comic patter all worked out, but it was meant for a very different audience. Early on in their gig, they said the fateful words "You know how it is when you're at a happy hour?" Silence. More silence. The two women waited with growing confusion. Somebody had to say it; I waited to give any colleague a chance to say it; nobody said it; so I yelled it out: "What's a happy hour?"

Happy hour? Who did they think they were dealing with? We were the weirdest of the weird. I'd never been to a happy hour, and I thought I was cooler than average in that room. These folks get off work and go straight home to prove theorems. The married guys still can't figure out how to ask their own wives out to dinner.

From a security perspective, the CCS gang is probably immune to all but one of the classical threats. Forget sex; until it is proven to help with proving theorems, it's a distraction. Forget money worries; these guys make enough not to have to think about money, which only distracts from proving theorems. Forget ideology; as long as the USA believes in math and more math, no problem. The one potential vulnerability might be ego.

The next summer, the same two women from Ft. Meade were back again for the annual security briefing. The first words out of their mouths: "We know none of you go to happy hours, so ..." Adapt and overcome. But then they got the treatment from us around the infamous computer security video.

This video depicts a number of scenarios, stopping periodically to ask the audience to make a selection from a list of choices about how to respond. For instance, suppose somebody you don't know says he wants to check your computer while you go to lunch. Should you agree (yes/no)? Of course, the whole auditorium starts to chant "yes". (OK, wrong answer. But of course. As they say, "obvious to the casual observer".) The movie asks whether it's ok to write your password on a sticky note and stick it to your monitor. Voices start yelling out the answer: "Yes", "Absolutely", "Write it bigger!", "Do it."

Two takeaways: Three cheers for esprit d 'corps. And stop treating us like we're stupid. Maybe a third takeaway: Aren't we the coolest nerds ever? (Actually no. If they showed the same video at MIT, they'd also get a barrage of paper airplanes, and they'd be the best paper airplanes ever, some circling for hours near the ceiling. But so many in the CCS auditorium went to Brown or Cornell or Berkeley that you can't expect too much from the group as a whole.)

The other big highlight of the security briefing is the swag giveaway. First, we are admonished not to put it around that we work at the distinguished IDA/CCS. Why make yourself a target of a foreign intelligence service or an ISIL sympathizer? Then the security staff proceed to hand out the goodies. [I believe it was ZACHARY who asked why they were contradicting themselves in such a way; that would have been in character. I cannot recall the mumbled answer.] The goodies vary from year to year, but the list includes CCS coffee mugs, CCS backpacks, CCS water bottles, and anything else that they can slap a CCS logo on for the public to (not?) see. Personally, I approve of the contradiction, since the swag is of pretty high quality. I don't drink coffee, but the CCS mug does a good job of holding pencils and pens, the CCS backpack is good for holding odds and bods in the back of my comfy little Honda Fit, and my daughter always has the CCS water bottle at her side at her job. Besides, all of it matches well with my NSA windbreaker. They say NSA used to stand for "No Such Agency". Now it stands for "Not So Anonymous". CCS? How about "Can't Conceal Secrets"? "Compiles Classified Swag"? "Carries Conspicuous Swag"?

That's not quite fair. Despite the hijinks when the official material is clumsy, everybody from the Director on down is dead serious about security. The important things get done right. The rest, though, can get a little goofy. CCS folks know the difference between life and death and Mickey Mouse. They react badly to Mickey and his friends. That's a good thing.

Life Outside while Inside

While NSA and CCS are keeping our minds occupied, they have to support us inside/outside types, since we summer people live our lives elsewhere.

As noted above, NSA Math Research supports its sabbatical visitors both financially and logistically. As a visiting academic, I continued to have my benefits and half my salary paid by my university. The Agency provided the other half of my salary. (Cheap at twice the price.)

NSA also provided a housing allowance (eventually increased to be on a par with military counterparts; see above) though no direct help with finding a place to live. My wife and I made a scouting trip and found a never-used/never-furnished apartment in Odenton, MD about half a mile outside the wire. Whereas I had the official status of "geographic bachelor" and could make do with a simple apartment, my fellow sabbatical visitor ANDREW was accompanied by wife and children and needed to find an entire house to rent.

I recall two perks that came with the apartment. One was free, no-charge interaction with Giant Voice. Giant Voice is a system of loudspeakers mounted all over Ft. Meade. One of its duties is to blast out the stirring bugle stylings of "Reveille" at dawn. The half-mile air gap between my air bed and Giant Voice was as nothing, and it was very "helpful" to be blasted awake every morning. In the evening, I received the more mellow treatment of "Taps", which never fails to move me.

In my previous sabbatical at the FAA, I lived near the Iwo Jima Memorial, so it was easy to wander into Arlington National Cemetery. The older I get, the mushier I get. Arlington National Cemetery, taps, the Memorial Wall at NSA all choke me up. Now on the outside, I still get to enjoy taps now and then. When visiting my company, I often stay at a hotel just outside the wire at Ft. Devens, MA. I can look out my window and see my old pal, Giant Voice, sitting on a telephone pole and poised to blast me out of bed. My dad mustered out of the Army at Ft. Devens in 1946, so I bet he had been more than happy to hear reveille there.

The other perk was access to the fitness facilities in our apartment complex. Ultimately, I found using this facility a little depressing. Almost every evening that I worked out, there was a woman there whose workout routine left me breathless from just watching. She told me she was an Army sergeant, which soothed my ego just a bit. Hey, I was just an oldish, overweight professor? It was a miracle that I was in the gym? Any excuse will do.

Part way through the sabbatical, I received a DoD identification card. This allowed me to take a shortcut through the gates of Ft. Meade proper. Ft. Meade is a large Army base, made even larger by later winning the BRAC (Base Realignment and Closing) sweepstakes, which closed several other operations and relocated them to Ft. Meade. NSA occupies a corner of the Ft. Meade space, but it is really a self-contained base-within-a-base. Without the DoD ID, I had to join the dense Route 32 traffic to drive around the Fort to get to the Agency. With the ID, I could drive through the Fort into NSA's back entrance.

That trip through the base was not only shorter but much more interesting. On the way in, I would drive by large drill fields where I would usually see lots of airmen doing their PT. On the way home, I would drive by the area where military police were training in hand-to-hand combat. I would slow down or stop to watch the action. At noon on weekdays, I would often drive over to the main chapel to attend Mass. (There was also an occasional mass in the Friedman auditorium at NSA, but the base chapel seemed more appropriate. After all, Friedman was where I'd attended my course in "Denial and Deception"; the juxtaposition seemed off key somehow.)

On perhaps too many occasions, I used my ID card to get through the wire and into the base Burger King, where I distinguished myself by probably being the only patron ever to read statistics books while chowing down on a Whopper. On weekends, I drove to the Cavalry Chapel for Sunday mass, then on to the ▮ building for more of my no-life work. One day, while cruising around the base, I was surprised to see that the base hosted Wiccan services every Saturday. Is the military again ahead of society in general? I guess Wiccan's who love the Constitution and can shoot straight are as good as any other soldier.

During my short time as a White Badge, the Agency housed me in an "executive apartment" during the summer. This was pretty cushy, with periodic maid service and fitness facilities. It also starting building up rewards points, which are a nice tax-free benefit. When I became a hated Green Badge and moved from NSA to CCS, housing continued to be provided in executive apartments until budgets shrank and we were downgraded to lower-end hotel rooms (still not so shabby compared to a posting to Kabul or Baghdad).

One of my more interesting summer housing assignments was my last executive apartment in Landover, MD. This was a rather opulent apartment, with a separate bedroom, living room, a large kitchen, a good internet connection, and a location smack in the middle of the approach path to Andrews Air Force Base. Having grown up near Westover AFB in Massachusetts, and having spent time at the

FAA, I am an "airplane spotter" rather in the mold of those famously odd British train spotters. Accordingly, I thought it was great to hear the "sound of freedom" rattling my windows with the noise from a wide range of heavy Air Force jets. I would listen to them on my aircraft scanner (the same one that HENRY once suspected was a spying device, to his eternal credit – he takes OPSEC seriously) and hear them lining up their approach on Fedex Stadium, home of the hated Washington Redskins and sitting just up the street looking like a crashed flying saucer. This apartment also gave me my first opportunity to "enjoy" the infamous Washington Beltway Traffic. I only had to jump onto the Beltway for a short distance before jumping off, but that was enough to make me appreciate my commute in upstate New York, where the biggest issues are avoiding collisions with kamikaze deer or, at certain times of year, turtles crossing the road to get to the Mohawk River.

The most interesting aspect of this particular housing assignment was racial. That summer, I got a good schooling on what it feels like to be a minority. The apartment was in a complex that abutted a Metro station. Except for me, SCOTT and a few other SCAMPers, most of the residents of the apartment complex were black. Whenever my neighbors would see me in the hallway, a look of surprise would flash across their faces. Ditto if I parked my pale self near the swimming pool or hauled my dumpy self to the fitness room. It was interesting to be conspicuous, since to white people I am eminently unremarkable.

Two days before I drove down to SCAMP, I sprained a knee while taking a quiet walk around my quiet suburban neighborhood. (Beware wet leaves!) As a result, I had to wear a knee brace and force myself to take a walk around the apartment complex to rehab my knee. Every morning I would put on raggedy shorts, raggedy t-shirt, beat up sneakers, ugly knee brace, and limp around, unshaven. All around me, the upwardly mobile black middle class would emerge to take the Metro down to DC to run the world. Each day was like an Easter parade. Everybody was dressed for success, groomed to the nines, and even smelled great. They would look at me and wonder who let the dumpy, gimpy white guy in. I was not keeping up the standard.

(They could not have known that, in the summer of 1968, I had been made an Honorary Black Man by my roommates in Kingston, NY, where we were all interns for IBM. I was a unanimous choice and remain very proud of that, though I was provided with no documentation by the gentlemen who thus judged me worthy. But in my heart, I wear a black badge beside the green.)

I got a rather warmer reception when I was dressed up for Mass. The closest Catholic church was St. Joseph's. When I showed up there for the first time, I saw that I was one of only three white faces (the other two were travelers driving south from Vermont). Everybody was very nice, but something memorably awkward happened. There seems to be a southern tradition of asking visitors to identify themselves in church. (I'd never encountered this very un-Yankee behavior, but I noticed it at the Ft. Meade chapel too.) When the priest asked all visitors to stand up (What!?), the three of us rose. (How could I not? I was well and surely busted.) The priest handed an altar boy a wireless mic and had him take it over to the Vermonters, who were way across the church. They introduced themselves, then the kid walked away with the mic. So there I was standing up on the other end of the church. Do I sit down? Do I keep standing? Too much pressure! After all, wasn't I supposed to be an inconspicuous Secret Person? I hung in and just stood there with a stupid grin on my face. Eventually, the priest spied me and got the altar boy to trek way over to my side of the church. By the time the kid finally reached me, I'd just about forgotten my name. But I managed an acceptable few words and finally sat down with some relief.

I continued to attend St. Joseph's and learned some other traditions. Besides singling out lost strangers, they made a big fuss about birthdays. I confess that when they asked everybody with a July birthday to stand, I pretended to not be what I really am – this time, nobody could tell I was lying by not standing. The priest and people seemed to really be interested in what day in July everybody's birthday was, where somebody was from, if they were visiting then whom they were visiting, and what they were doing here in the first place. Too much for me: Way too Southern. (I grew up with the New England rule: You can say hello to your abutters after a five year waiting period, then things can move on from there at a dignified pace.) I salved my conscience by telling myself it was ok because I was not really allowed to say why I was in Maryland, and I was trained, at minor public expense, in denial and deception (even at Mass?).

One Mass that I will never forget commemorated the founding of St. Joseph's. A little old lady was asked to speak because she had been at the very first mass celebrated there when she was just a little girl. I will always remember with shame her story: all the black Catholics had to wait to receive Holy Communion until all the white Catholics had gone first. Unimaginable.

During subsequent summers, SCAMP budgets could not support the luxury of the executive lifestyle. CCS even had to trim the official government per diem reimbursement in order to afford reasonable housing. My housing switched to a motel near Annapolis, MD. This arrangement had certain advantages: A goodly number of us were housed there, so it was easy to meet, say, a secret ███ nerd over breakfast; there were some good restaurants within walking distance, the beautiful Quiet Waters Park was a short drive away, and we drove counter-flow to CCS so the traffic was bearable. We were technically in Parole, MD (some significance to that?) but beautiful Annapolis was close by and provided many ways to spend money if so inclined, especially on tourist trinkets and crab cakes. My church routine was rather different, in that St. Mary's was a very rich, very white parish, where the Sunday collection equaled the monthly collection at my church in upstate New York. (A bit of amusement: my Garmin GPS was more interested in geography than theology. As I got close to the church, it would announce that I was arriving not at St. Mary's but at "Street Mary's". Funny but incorrect: this church was definitely not "street".)

Next Man Up

We'll end this tale of my adventure on the dark side of the moon with the image of me on my knees in prayer at Street Mary's. That's not a bad posture, after so many hours reading the "NSA Daily" and seeing a world of threat invisible to most of us outside. I find it as much depressing as threatening, in part because I see no end to it.

While I have been darting in and out of the IC, Al Qa'Ida was weakened and Bin Laden killed, wars more or less wound down in Iraq and Afghanistan, and Iran's acquisition of The Bomb was postponed. But at the same time, Daesh rampaged, Russia invaded my grandfather's Ukraine, North Korea got more dangerous in both the nuclear and cyber realms, China loomed, and America was subject to a barrage of cyberattacks big and small. Meanwhile, Edward Snowden and his apologists caused many Americans to doubt the integrity and value of the NSA at a time when it is challenged as never before to protect our national security.

My involvement in this vale of tears began with the NSA sabbatical program. I believe the Agency got some mileage out of my time inside. They got even more out of the other sabbatical professor that year, ANDREW, who committed to full-time service, and the Agency got a two-fer when his son CALVIN also signed up. The sabbatical program is a weapon.

Perhaps this book will become my largest contribution to the Agency. If it sheds some light on the dark side of the moon, the men and women who work there will have more support for their invisible but important labors on our behalf. If it stirs some interest among other professors to add their brains to the fight, that too will help. It is obvious to the casual observer that there are more serious problems than we have serious brains to throw against them.

I hope someone, man or woman, will soon take my place at our secret base on the dark side of the moon.

CPSIA information can be obtained
at www.ICGtesting.com
Printed in the USA
BVHW051409280423
663226BV00012B/1144

9 781629 528724